S0-BHX-933

The State of the Rio Grande/Río Bravo

The Texas/Mexico border stretches nearly 1,200 miles and is home to more than two and a half million people — a border both separated and united by water. This new study brings together in a single volume a wealth of data, previously available only in government publications and specialized works, relating to water stocks, uses, and institutions in order to isolate the facts and trends that underlie hydrologic engineering, and economic realities for the region's future.

The book addresses water quantity and quality issues along the Rio Grande/Río Bravo watershed. It considers groundwater as well as surface water, the relationship of water supply to disease, and the impact of rapid demographic change on water supply. Original research has been undertaken to identify and catalog types of water-management institutions and treatment systems on both sides of the border.

Despite the ongoing availability of water-resource statistics, there has been a lack of usable information on policy-related matters. Eaton and Andersen's study now provides that information, in a format that will be of as much value to urban planners and legal specialists as to hydrologists.

David J. Eaton is Professor of Public Affairs and Geography at the Lyndon B. Johnson School of Public Affairs, the University of Texas at Austin. John M. Andersen is International Economist in the International Trade Administration, U.S. Department of Commerce, Washington, D.C.

The State of the
Rio Grande/Río Bravo:
A Study of
Water Resource Issues
Along the
Texas/Mexico Border

DAVID J. EATON
JOHN M. ANDERSEN

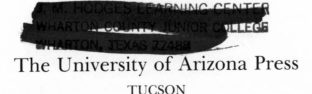
The University of Arizona Press
TUCSON

THE UNIVERSITY OF ARIZONA PRESS
Copyright © 1987
The Arizona Board of Regents
All Rights Reserved

This book was set in 10/12 VIP Times Roman.
Manufactured in the U.S.A.

Library of Congress Cataloging-in-Publication Data
Eaton, David J.
The state of the Rio Grande/Río Bravo.
(Profmex monograph series)
Bibliography: p.
Includes index.
1. Water-supply — Rio Grande Watershed. 2. Water
consumption — Rio Grande Watershed. 3. Water quality
management — Rio Grande Watershed. I. Andersen, John M.
(John Michael) II. Title. III. Series.
TD225.R53E28 1987 333.91'009764'4 87-10748
ISBN 0-8165-0990-5 (alk. paper)

British Library Cataloguing in Publication data are available.

Table of Contents

List of Figures

Lists of Tables

Preface

The relationship between water and people vacillates between poetic wonder and idle disinterest. Water, embodied in rivers, lakes, and oceans, has been recognized for its mystical qualities in religion, legend, and folklore. There have been partings of the sea, prehistoric creatures of lagoons, and even the salty tears of La Llorona as she wanders the river. Water has sanctified, purified, performed acts of justice -- or injustice -- and provided humanity and literature with a rich metaphor for life and all that is vital.

At the same time, water has been regarded as boundless, cheap, and dispensable. Adam Smith in *The Wealth of Nations* in 1776 presented the famous diamond-water paradox:

> Nothing is more useful than water: but it will
> purchase scarce any thing. . . . A diamond,
> on the contrary, has scarce any value in use,
> but a very great quantity of other goods may
> frequently be had in exchange for it.

The reason, said Smith, for this apparent contradiction could be explained by the two types of value which express themselves in a good: *value in use* and *value in exchange*. Water has great value in use but little in exchange, so the argument goes.

Two hundred years later, the value in use of water has increased with the complexity of society. Water refreshes, cleanses, irrigates, and drives water-powered manufacture. Water is the universal solvent, the cooling fluid for electrical energy generation, and the cleansing agent for human and industrial waste. Water is used for drilling, mining, injecting, and displacing energy and minerals from beneath the earth's surface. Water turns generators, shelters boat docks, and provides a surface and medium for human recreation. Water feeds gardens, nourishes estuaries, and quenches thirst. Water, a simple alliance of two elements, fastens our lives, our economics, and our resources together in an exceedingly complex fashion.

Water has a tremendous value in use and thus under scarce conditions has undiminishing value in exchange. Water, as it always has, determines and sustains the progress and accomplishments of a civilized society. What we have known from tale and tradition, therefore, is truer now than ever before.

This report began as an effort to measure, understand, and then recommend policies for the human uses of the Rio Grande/Río Bravo region's water. Over the course of research, the border region has provided the participants with a superior challenge.

On the cover map, the territory is less than twelve inches long; in reality, the Texas-Mexico border stretches nearly twelve hundred miles. The region appears sparse in population, yet nearly 2.7 million persons occupy the frontier and more will reside here in two decades. The region seems barren and lifeless, but over 90 percent of the annual water withdrawals are for irrigation. Water is neither all murky nor all clear; the 58 water quality monitoring stations register 250 ingredients, some hazardous and some essential to the river's life and to the people who live nearby. And as this project has proceeded, the layers enveloping our original perception of this region have peeled away to reveal an immensely complex and illusory frontier.

The challenge of this research has been to try to isolate -- from a large volume of data -- the facts and trends that underlie hydrologic, engineering, economic, and institutional realities. Chapter One reports on the region's surface water availability, including both United States and Mexican tributaries. Chapter Two discusses the water quality along the Rio Grande/Río Bravo, including the standards, monitoring stations, and the contaminants of the basin. Chapter Three provides special information regarding regional groundwater availability and quality. Chapter Four presents a population profile of the borderlands and projections of that population to the year 2000. Chapter Five is a comprehensive survey of municipal, irrigation, steam-electric, manufacturing, and mining demands along the Rio Grande/Río Bravo. Chapter Six discusses the types and methods of water treatment in the border region. Chapter Seven contains information on wastewater treatment processes. Chapter Eight includes information on the regional water-related health problems. Chapter Nine surveys the methods and results of water demand forecasting for the region, with a view toward the next fifty years.

This report describes the water quality and quantity issues that separate and unite the border between Texas and Mexico. Project members do not consciously seek to influence Texan or Mexican choices regarding the basin's water resource problems. Given limited time and resources, we have tried to state the facts in a manner that will be perceived as fair by persons with a wide variety of policy perspectives. There are, however, several topics of nonpartisan concern and significance which merit more detailed consideration.

One matter could be termed "the simultaneous feast and famine" of water resource statistics concerning the Rio Grande/Río Bravo basin. Institutions in the basin -- including the IBWC, state agencies of Texas, and Mexican government ministries -- publish a plethora of numbers relating to water quantity, water quality, and water withdrawals. This feast of available numbers obscures an absence of useful information on policy-relevant matters, including *who uses how much water from which source for what purposes.*

Project members have made an effort to organize these disparate water-related facts. We were not able to do more than survey Mexican water withdrawal data and had neither the time nor the resources to obtain and integrate Texas water permit information. Our impression is that coordination of water data collection and reporting by agencies on both sides of the border could greatly assist water resources planning.

A second concern is that water resources management in the basin suffers because *the persons who make decisions may differ from those persons who implement them; neither set may consist of people whose vital interests are affected by these policies.* Both Mexico and the United States reserve substantial powers to the federal governments. Mexico's water quantity and quality policies are to a large degree established by federal agencies in Mexico City and implemented by civil servants resident in the affected region but responsible to these federal institutions. Although surface and drinking water quality standards are established by federal agencies in the United States, implementation responsibilities reside with the states. Texas, for example, has its own surface water quality and drinking water hygiene standards. Texas also operates a water rights permit program regulating surface water withdrawals. In the Rio Grande/Río Bravo basin, water withdrawals are administered by the IBWC, under water quantity and quality understandings reached by the governments of Mexico and the United States through binational treaties. Although the staff of all of these institutions are familiar with many of the water issues of the basin, they remain civil servants of their respective state or national institutions. In short, the decision-making powers reside in Mexico City, D.F., Washington, D.C., or Austin. Policies are implemented by persons whose primary loyalties lie outside the basin.

If there is one generalization that can be made from the variety of data presented in this volume, it is that water resource problems along the Texas/Mexico border affect specific sites. Although the pattern of water consumption or return flows in one part of the basin does affect the other parts, problem areas reflect proximate practices. The scale of the region is so vast and the sources of inflow are so numerous that water availability problems occur in areas with heavy irrigation and municipal water demands from both sides of the border. Water quality problem areas are often adjacent to sister cities.

One could easily marshall such facts into an argument for permitting local water resource institutions some autonomy for reaching site-specific agreements which implement accepted binational, national, or state policies. For example, an organization of groundwater users should be able to establish self-enforced withdrawal limits to conserve groundwater in areas suffering from depletion or saline infiltration. Why shouldn't sister cities across the border be encouraged to jointly plan, build, and operate water supply and waste treatment facilities? The sensitivity of each nation to its sovereignty should not blind leaders to the advantages of delegating limited implementation powers to those most directly affected.

There are many other issues that are beyond the scope of our work but which should be addressed. For example, water resources development requires money, whether to convert inefficient irrigation canals to drip-tube technology or to plan/build/operate/maintain sophisticated water and waste treatment plants. Training water resources personnel is another topic that could be discussed, or perhaps the level of technology appropriate for managing water quality in a context of two sets of national standards.

Shared river and groundwater resources can either bind neighbors together in cooperative development or undermine peaceful relations. The challenge for water resource managers within the Rio Grande/Rio Bravo basin is how to make efficient use of a scarce, shared resource in an uncertain future. This task is complicated because the requests for water and the costs of providing it may tax the two nations' willingness to pay, as well as their technical capabilities.

Acknowledgments

The research for this book was conducted by thirteen graduate students and four faculty members in the Lyndon Baines Johnson School of Public Affairs of The University of Texas at Austin. The graduate students include John Michael Andersen, Jeffrey R. Cole, Mary Elaine Eakes, Leslie J. Friedlander, Eduardo Ganem-Musi, James Kevin Gatz, Richard J. Holden, Kim Blanton Lemon, Alfonso Ortiz Nunez, Bill Paschall, Deborah Ann Sagen, Pablo Salcido, and Thomas Troegel. Professors David J. Eaton, Richard S. Howe, Milton Jamail, and Pedro Martinez-Pereda were the faculty.

This research was supported by grants from the Middle Rio Grande Development Council, the Lyndon Baines Johnson Foundation, and the University Research Institute of The University of Texas at Austin. In addition, many individuals and organizations contributed time, information, ideas, and encouragement.

The Policy Research Project members express appreciation to the following individuals from the United States for their assistance in the preparation of this report: Norman Alford, Shelby Anderson, Victor Arnold, Gerald Baum, Shaul Ben-David, F. G. Bloodworth, Blair Bower, Dave Buzan, Don Dawkins, Marilyn Duncan, Irving Fox, Rosario Ganem, Robin Greenley, Barbara Griffen, John Gronouski, Herbert Grubb, Joe Haggard, Niles Hansen, Dan Hardin, Joe Harris, Gerald Higgins, William H. Hoffman, Robert Johnson, Donna Jones, Ken Kramer, Miriam Lemay, Randy Lyon, Harrold Patterson, Mike Personette, Gerard Rohlich, Stanley Ross, Marilou Salmon, Alan Siles, Chandler Stolp, Nancy Sugg, Zdenko Tomasic, Ben Turner, David Warner, Sidney Weintraub, and Mickey Wright.

We are also grateful to many individuals from Mexico who gave freely of their time and information: Jorge Aguirre Martinez, Enrique Campos Lopez, Ariel Cano Vicario, Juan Antonio Castillo, Enrique Dau Flores, Fernando Gonzalez Villarreal, Salvador Hernandez Pacheco, Francisco Malagamba, Carlos Malpica Flores, and Francisco Tello Vasconcelos.

In addition, we wish to thank the following institutions in the United States and Mexico for their cooperation: Computation Center of The University of Texas at Austin, The International Boundary and Water Commission, Population Research Center of The University of Texas at Austin, Sierra Club (Lone Star Chapter), Texas Department of Health, Texas Department of Water Resources, Texas State Archives, Centro de Estudios Fronterizos del Norte de México, Centro de Investigaciones en Química Aplicada, Secretaría de Agricultura y Recursos Hidráulicos, Secretaría de Desarrollo Urbano y Ecología, and Secretaría de Salubridad y Asistencia.

We appreciate this assistance from many persons and institutions; any remaining errors or omissions are ours.

David J. Eaton
John Michael Andersen

CHAPTER 1
Surface Water Availability

The Rio Grande/Río Bravo basin straddling Texas and Mexico has been the land of the Pueblo Indians, home to Spanish settlers, and an area where people sought refuge from the law. Today the river basin between El Paso/Juárez and Brownsville/Matamoros unites the United States and Mexico, and its waters still mean life to the region's 2.7 million inhabitants.

The Rio Grande/Río Bravo basin extends 1,885 miles (3,033 km) from southern Colorado through middle New Mexico to the Gulf of Mexico near Brownsville/Matamoros (see Figure 1.1) (1). This region includes portions of Colorado, New Mexico, and Texas (in the United States) and of Chihuahua, Coahuila, Nuevo León, and Tamaulipas (in Mexico). With a total land area of 355,500 square miles (920,389 sq km) (2), it is equal to 11 percent of the continental United States and to 44 percent of the land area of Mexico (Table 1.1) (3). Of approximately 135,500 square miles (350,809 sq km) in the United States and 199,100 square miles (515,469 sq km) in Mexico, about 88,968 U.S. square miles (230,338 sq km) and 87,365 Mexican square miles (226,187 sq km) contribute streamflow to the river (4); the remaining area drains internally without runoff. Texas's portion of the basin is approximately 48,300 square miles (125,048 sq km), although 38,800 square miles (100,453 sq km) contribute streamflow (5).

This chapter describes the river, the region, and the available surface water resources. The first section briefly introduces the Rio Grande/Río Bravo itself. It describes the river's course, its major tributaries and reservoirs, and the geology of the region. A second section evaluates the river's flow in terms of the basin's extremely varied climate; the northern portions are dry and arid while the southern reaches

1

Figure 1.1

Rio Grande/Río Bravo Basin

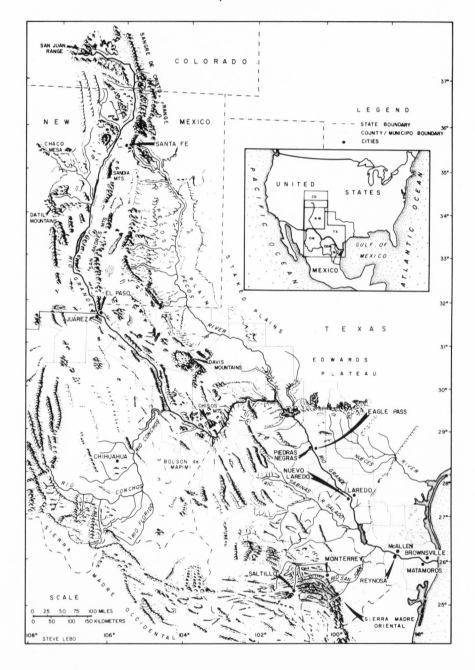

are wet and tropical. Such weather-related measures as precipitation, temperature, evaporation, relative humidity, and wind movement throughout the Texas/Mexico border are included. A third section assesses where and how water availability is affected by tributaries, canals, diversions for irrigation, and water stored along the river's course.

THE COURSE OF THE RIO GRANDE/RÍO BRAVO

The Rio Grande/Río Bravo's headwaters emerge twelve thousand feet above sea level in the San Juan Mountains of southwestern Colorado (6). As the river flows southward through central New Mexico, it passes through a plain bordered on the east by the Sangre de Cristo Mountains and on the west by the San Juan and Jiménez mountains (7). The river enters Texas twenty miles (32 km) northwest of El Paso and unites the United States and Mexico from El Paso/Juárez to the Gulf of Mexico near Brownsville/Matamoros (Figure 1.2). For the 1,250 miles (2,012 km) between El Paso and the Gulf, the river flows through three sections or reaches, which will be referred to frequently throughout this volume (8). Along the Mexico/Texas border there is a dry, arid region (upper reach), followed by a second stretch (middle reach), where the river is reborn by powerful tributaries and nourished by two international reservoirs. Much of the flow of the third section (lower reach) is extracted for irrigation. The Rio Grande/Río Bravo eventually deposits its water into the Gulf of Mexico twenty-three miles (37 km) east of Brownsville/Matamoros.

In the upper reach, from the New Mexico border to a point just above the Río Conchos, the river flows through a relatively flat, low basin (9). At Fort Quitman it enters a series of valleys and canyons. For the 212 miles (341 km) below, its current is reduced to a trickle (10).

As the river enters the middle reach near Presidio/Ojinaga, it merges with the Río Conchos (at river mile 963.7 or km 1,551), which regenerates the Rio Grande/Río Bravo. The Conchos originates in the Tarahumara Mountains of Chihuahua and Durango, Mexico. The perennial flow benefits from the relatively heavy precipitation and snowmelt from the high elevations of the region (11). Five tributaries in the Río Conchos basin--the Río Florido, Río San Pedro, Río Bachimba, Río Chuviscar, and Río Parral--continually feed this Mexican river. Five reservoirs store the water of the Río Conchos basin: the San Gabriel, Hidalgo del Parrel, La Boquilla, Madero, and Luis León reservoirs. It is the waters of the Conchos which make up the Rio

Figure 1.2

The Texas/Mexico Portion of the Rio Grande/Río Bravo Basin

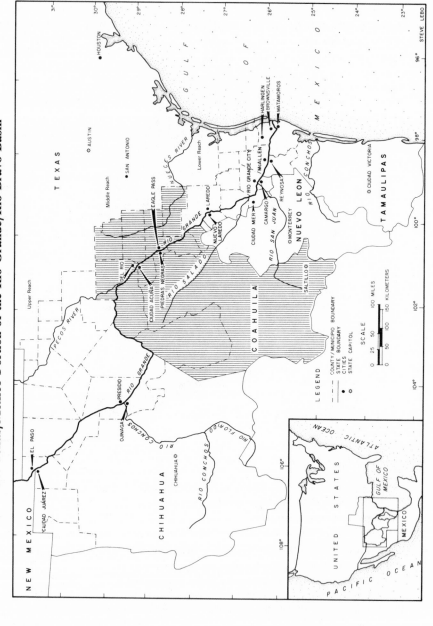

Grande/Río Bravo as it passes through a wide valley toward the canyons of the Big Bend area. The dramatic topography of the Big Bend wilderness and National Park divides the river into three canyons: the Santa Elena, Mariscal, and Boquillas. The deepest of these canyons is 1,600 feet (490 m) below the surrounding terrain, and the longest is 25 miles (40 km) (12).

Beyond Big Bend, the Rio Grande/Río Bravo continues to flow northeast and east through 135 miles (217 km) of limestone canyons and mountains until it reaches Langtry, Texas (13). Along this stretch, both Alamito and Terlingua creeks feed their water to the river from the north. Below Langtry, the river's rate of fall is greatly reduced; it meanders through occasional low hills and narrow valleys until it reaches the Amistad Reservoir (14). Amistad Reservoir, the first of two major reservoirs on the river, is located twelve miles (19 km) northwest of Del Río, Texas, and Ciudad Acuña, Coahuila.

The Pecos and Devils rivers also discharge their water into the Amistad Reservoir and thus into the Rio Grande/Río Bravo. The Pecos's headwaters begin in the Sangre de Cristo Mountains northwest of Santa Fe, New Mexico. Its major tributaries in New Mexico are the Río Peñasco and Alamogordo creeks; the Delaware, Toyah, and Independence creeks are its main sources in Texas. Lake McMillan and Lake Sumner in New Mexico and Red Bluff Reservoir near the Texas/New Mexico border regulate the Pecos's flow. Tributaries of the Devils River include Big Satan Creek, Rough Canyon, and Dolan Springs in Texas (15).

As the river continues its course south of Amistad Dam, its water is utilized by the major population centers of Del Río/Ciudad Acuña and Eagle Pass/Piedras Negras. It is in this stretch that four Mexican rivers--the Río San Diego, Río San Rodrigo, Río Escondido, and Río Las Vacas--deposit their water in the Rio Grande/Río Bravo (16). The San Diego and San Rodrigo begin in the del Burro Mountains of the Mexican state of Coahuila, flowing eastward toward the Río Bravo (17). The Escondido also emerges in the del Burro Mountains and joins the Río Bravo just four miles downstream from Piedras Negras (18).

Farther south, eighty miles (128 km) from the cities of Laredo/Nuevo Laredo, the river's water is impounded in the second and last international reservoir in the basin--Falcon Reservoir. The main contributor to the reservoir, and thus to the river at this point, is the Mexican Río Salado. The Salado drainage basin includes the Sabinas, Salados de la Nadadores, Sabinas Hidalgo River, Cameron Creek, Venustiano Carranza Reservoir, and the Don R. Martin irrigation district, all within the Mexican states of Coahuila and Nuevo León (19). For the next 250 miles (402 km) downstream from the dam the basin is very narrow and flat. About twelve miles (19 km) downstream from Falcon Dam, the Río Alamo, another Mexican tributary, deposits its

water in the Rio Grande/Río Bravo. The Macho and Lajitas creeks and
the Río Sosa flow into the Río Alamo. Tributaries to the Río Sosa are
the Reyes, Yrias, and San Jeronimo creeks (20).

The Rio Grande/Río Bravo is again fed by water from a Mexican
river thirty-six miles (58 km) downstream from Falcon Reservoir. This
source is the Río San Juan Catarinas which in turn is fed by the Río
Sabinas, Río Pesqueria, Río Santa, Río Ramos, Río Pilón, and Mohinas
Creek (21). Two major Mexican reservoirs in this basin are the Rodrigo
Gómez at La Boca and the Marte R. Gómez at El Azucar (22). The
proximity of the Gulf of Mexico exposes this lowest portion of the basin
to influences of gulf-related weather activity (23). As the Rio
Grande/Río Bravo's water heads toward the Gulf, much of its flow is
diverted for irrigation.

In all, there are approximately thirty measurable sources of
streamflow to the Rio Grande/Río Bravo between El Paso/Juárez and
the Gulf of Mexico, nine major diversions, and two reservoirs (24).
Figure 1.3 details flow deposits and withdrawals along the river as
measured by the International Boundary and Water Commission
(IBWC). The vertical axis represents the Rio Grande/Río Bravo. The
area on the left represents the United States, and that on the right represents Mexico. Deposits are represented by arrows leading to the vertical line and withdrawals are delineated by arrows leading away from
the river. The numbers indicate the volume of water contributed and
withdrawn at each point both in 1980 and over a multi-year period.

As indicated in Figure 1.3, 353,983 acre-feet (436.6 million cu m)
of water flowed in the Rio Grande/Río Bravo from New Mexico into
Texas in 1980. If this is considered the base volume of flow, the inflows
and outflows downstream from El Paso can be viewed as adding to or
taking water from this amount. The major contributors include the
Conchos, Pecos, Devils, San Rodrigo, Alamo, and San Juan tributaries.
A total of 2,771,753 acre-feet (3.417 billion cu m) were measured as inflow to the river in 1980. River diversions in 1980 totaled over
3,964,690.5 acre-feet (4.89 billion cu m). Diversions take place primarily
in three general areas--at El Paso/Juárez, south of Amistad Dam, and
south of Falcon Reservoir.

The water balance along the river is monitored and allocated by the
International Boundary and Water Commission according to the for-
mulas outlined in the Rio Grande compact of 1939 and two
U.S./Mexico treaties, one signed in 1906 and the other in 1944 (25). The
earlier treaty addresses the delivery of water to Mexico from the United
States in the segment from El Paso to Fort Quitman. The later agreement deals with allocation from Fort Quitman to the Gulf of Mexico.
It also empowers the IBWC to enforce the treaties' articles through flow
monitoring to ensure that river flow allocation to each country is in
accordance with the international agreements.

Figure 1.3

Inflow and Outflow Along the Rio Grande/Río Bravo

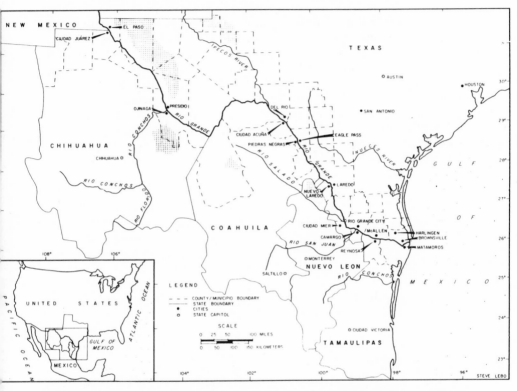

Source: International Boundary and Water Commission of the United States and Mexico--United States Section, *Flow of the Rio Grande and Related Data, 1980*, Water Bulletin 50 (El Paso: IBWC, 1980), pp. 7-78. *Flow averages presented in this figure were modified by the IBWC to include only the years occurring after completion of major projects below which the flow of the Rio Grande/Río Bravo or a primary tributary was affected, or at a later date when records were made available. The revisions are based on the completion of Caballo Dam (1938), irrigation works on the Rio Conchos and its tributaries (1947), Falcon Dam (1953), and Amistad Dam and Luis L. Leon Dam (1968).

The amount of water present in a river basin at any given time is dependent upon several interrelated factors. Surface water streamflow and groundwater volume are the two basic components. Streamflow includes both surface water runoff and groundwater flow that emerge at the earth's surface. Surface water runoff is related to precipitation, evapotranspiration, and infiltration; precipitation minus evapotranspiration and net infiltration equal surface water runoff. The amount of natural runoff is dependent on the area of the basin examined, the soil composition of the area, the type of vegetation present, the depth of the water table, the slope of the land surface, and the distribution of precipitation. The second component, groundwater volume, also depends on precipitation. Aquifers are recharged by rainfall and surface water runoff that infiltrates through the soil to the water table. They are also supplied by infiltration or percolation from surface water bodies, such as lakes, reservoirs, and rivers. The amount of water reaching an aquifer from the surface is dependent on the soil type and the climate. The water level in any groundwater reservoir is more or less related to atmospheric and surface water (26).

CLIMATE

The climate of the Rio Grande/Río Bravo basin has a direct influence on the river. Precipitation supplies the runoff that eventually constitutes river flow. Temperature, wind movement, and humidity affect the basin's evaporation rate, and thus the portion of precipitation that becomes flow. The climate supplies (through precipitation) and takes (through evapotranspiration) both flow and potential flow from the basin. This section will consider several climatic factors of the basin-- the levels of precipitation, temperature, humidity, evaporation, and transpiration.

A geographer would classify the Rio Grande/Río Bravo basin as a tropical or subtropical desert (27). The region is a warm, windy, and predominantly sunny corner of the globe. Such generalizations, however, are misleading, as no one climatic type accurately describes the total length of the basin. The climate of the northern portions differs from that of the southern reaches. The northern section's characteristics include desert land, scarce water resources, and high temperatures; in the south, residents are blessed with high precipitation and a more tropical climate. Table 1.2 lists climatic extremes recorded in the Rio Grande/Río Bravo basin.

Precipitation

Although some snow falls in the higher elevations of Chihuahua, Mexico, essentially all the precipitation in the Texas/Mexico portion of the basin is in the form of rainfall (28). Precipitation is relatively light in the western portions of the basin; it becomes heavier approaching the Gulf of Mexico. The yearly precipitation totals range from 10 inches (25.4 cm) near El Paso/Juárez to approximately 23 inches (58.42 cm) in the far eastern and downstream regions of the basin near Río Grande City and Matamoros (29). There is a more-or-less continuous increase from the northern to the far southern segments of the basin.

Precipitation does not occur uniformly throughout the year, nor does it occur equally in all segments of the basin. The "rainy season," if in fact such a season exists, is limited in the western sections to August through September (30). Progressing eastward along the river, the Gulf's climate gradually lengthens the rainy period. Beginning at Foster Ranch and continuing with each segment farther downstream, the period April through October contains the wettest months (31). Rainfall occurs chiefly in the form of brief but heavy summer thunderstorms. Rain falling during November through March typically occurs as light showers (32). Dry periods of several months' duration are not unusual, especially in western sections of the basin (33).

Temperature

Like precipitation, temperatures vary greatly from the western to eastern segments of the basin. Average annual temperatures of the northern portions are generally less than those of the southern reaches. The basin as a whole experiences warm summers and relatively cool winters. Both summer and winter temperatures are higher in the southern areas (34).

Although mean monthly air temperatures vary no more than fifteen degrees between sites in the basin, the maximum and minimum temperatures vary widely at any one site, from day to day, or from month to month. July and August are the warmest months, with mean temperatures ranging from 82 to 92 degrees Fahrenheit (27 to 33 degrees

Celsius) (35). December and January are the coldest months; mean temperatures range from 49 to near 60 degrees Fahrenheit (9 to near 15 degrees Celsius) (36). These figures are derived from IBWC measurements at seventeen sites, three in the United States and fourteen in Mexico (36).

Wind Speed

Wind speed can affect water availability, as high wind velocity is one factor in a high evaporation rate. The IBWC measures wind speed at four sites, all located in the United States (37). Table 1.3 lists the 1980 mean and multi-year average monthly wind speeds for all four monitoring stations. April through August is typically the windiest time of the year, while November through January is the most tranquil.

Humidity

Relative humidity also influences water availability by affecting the evaporation rate in the basin. Humidity is measured by the IBWC at three stations in Texas and one in Mexico (38). Measurements taken at these sites are included as Table 1.4.

Evaporation

Evaporation, the movement of moisture from the soil and water surfaces to the atmosphere, varies with wind velocity, temperature, and humidity. In general, high evaporation rates are associated with high wind speeds, temperatures, and relative humidity.

The IBWC operates nineteen observation stations to monitor evaporation in the basin. Eight of these stations are in Texas and the remaining ten are in various Mexican states (39). Average annual evaporation rates vary widely from station to station. The evaporation rate tends to decline from west to east in the Texas region, with the exception of the areas near the Amistad and Falcon reservoirs. Evaporation rates recorded in the Mexican region are higher and more constant than those recorded in the United States. The highest evaporation rates at each IBWC station are associated with the period of May through October.

The significance of the evaporation measurements are clarified by comparing the magnitudes of evaporation and precipitation (see Tables 1.5 and 1.6). At each location the evaporation rate is much greater than precipitation. For example, at Amistad Dam the average precipitation is 19.19 inches (48.74 cm) and the expected annual evaporation rate is 105.06 inches (266.85 cm), for a net annual water loss of 85-87 inches (40).

Transpiration

The transformation of water in the soil to water vapor in the atmosphere through plants is known as transpiration. Plants utilize water to move nutrients from the soil to their tissues and for chemical reactions carried out within their cells (41). The water that travels from the roots toward the outer surfaces of the plant is then discharged into the atmosphere as vapor. A high transpiration rate indicates a high rate of water absorption by the roots of the plant. Evaporation and transpiration rates are highly correlated; humidity, temperature, and wind movement influence transpiration (42). The term evapotranspiration is often used to describe the combined evaporation and transpiration rates.

Species of plants known as phreatophytes have roots which draw water from water-saturated areas of groundwater aquifers. Saltcedar, cottonwood, and willow are examples of this type of plant found in the Rio Grande/Río Bravo basin. They commonly grow along the shorelines of reservoirs or creeks and in other places where the water table may be near the surface (43). Saltcedar is prevalent along the Rio Grande/Río Bravo, especially upstream from the confluence of the Pecos River. A 1967 survey by the U.S. Soil Conservation Service showed about 105,200 acres (42,570 ha) of saltcedar in the Rio Grande/Río Bravo basin upstream from Amistad Reservoir (44). The U.S. Soil Conservation Service has estimated that these phreatophytes lose 3.26 acre-feet (4,021 cu m) of water per acre of saltcedar, for a cumulative water loss in the basin of 343,000 acre-feet (423.0 million cu m) through evapotranspiration every year (45).

SURFACE WATER AVAILABILITY AND STORAGE

The volume of the Rio Grande/Río Bravo between El Paso/Juárez and the Gulf of Mexico varies over time and space. The westernmost segments often contain critically low amounts of river flow. Lower Rio Grande Valley segments can have excess flow. The Amistad and Falcon reservoirs are operated to manage and maintain the flow in the eastern river segments.

Figure 1.4 illustrates monthly minimum, maximum, and total annual flow, at 22 gauging stations along the Rio Grande/Río Bravo for 1980 and a multi-year period. Flow rates for 1980 ranged from a high of 2,635,716 acre-feet (3.25 billion cu m) per year below Falcon Dam to a low of 1,384.4 acre-feet (1.7 million cu m) near Clint/San Agustín.

Nineteen major reservoirs can be found in the Mexico/Texas region of the basin. Three projects are in Texas and fourteen are distributed in the Mexican states of Chihuahua (eight), Coahuila (two), Nuevo León (one), and Tamaulipas (three). Two international reservoirs in the region, the Amistad Reservoir near Del Río/Ciudad Acuña and the Falcon reservoir near Zapata/Ciudad Guerrero, are owned jointly by the United States and Mexico and managed by the IBWC (46). Figure 1.5 shows each reservoir's location.

Reservoirs influence the mass water balance of the basin in several ways. Reservoirs withhold and thus control downstream flow; captured streamflows can be released in accordance with downstream needs. The IBWC uses both Amistad and Falcon to control and manage water resources for the lower Rio Grande/Río Bravo in this fashion. Percolation and evaporation from reservoirs reduce the volume of surface water. As water is stored, gravity pulls some fraction of the water down into underground aquifers and the climate induces some fraction of the water into the atmosphere.

Underground water storage also exists in the basin (see Chapter Two). Subsurface water is influenced by the environment through the types of vegetation and soil in the area and evapotranspiration. Groundwater seepage can reenter the surface in the form of springs or infiltration to rivers from adjacent banks. One important factor of groundwater is that it can move horizontally below the ground surface. Heavy pumping of water from an aquifer or a natural discharge of an

(continued on page 16)

Figure 1.4a

**Multi-Year Maximum, Minimum and Average Flow of the Rio
Grande/Río Bravo
(acre-feet)**

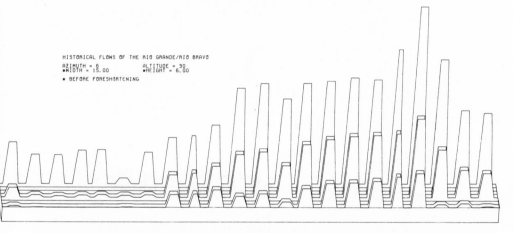

Source: International Boundary and Water Commission of the United
States and Mexico--United States Section, *Flow of the Rio Grande and
Related Data: 1980*, Water Bulletin 50 (El Paso: IBWC, 1980), pp.
7-83.

The figure illustrates multi-year flow levels in the main stem of the river
ranging from zero to more than 6,619,700 acre-feet.

The three rows represent the maximum (back row), median (middle
row), and minimum (front row) flow levels measured at 22 monitoring
stations. The far left corner corresponds to the New Mexico border and
the right to the Gulf of Mexico. The vertical scale correctly portrays in
perspective the relative flow peaks corresponding to the areas where
U.S. and Mexican tributaries enter the river. The horizontal axis is only
a rough approximation of relative horizontal distances along the length
of the Rio Grande/Río Bravo.

Figure 1.4b

**1980 Maximum, Minimum and Average Flow
of the Rio Grande/Río Bravo
(acre-feet)**

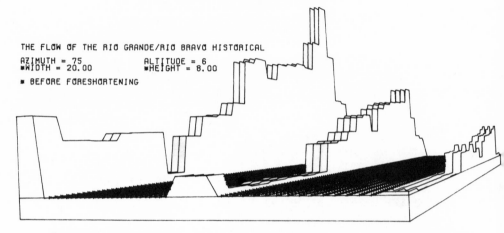

THE FLOW OF THE RIO GRANDE/RIO BRAVO HISTORICAL

AZIMUTH = 75   ALTITUDE = 6
*WIDTH = 20.00   *HEIGHT = 8.00

* BEFORE FORESHORTENING

Source: International Boundary Water Commission of the United
States and Mexico--United States Section, *Flow of the Rio Grande and
Related Data: 1980*, Water Bulletin 50 (El Paso: IBWC, 1980), pp.
7-83

The figure illustrates 1980 flow levels in the main stream of the river,
ranging from zero to more than 2,635,716 acre-feet.

The three rows represent the maximum (back row), median (middle
row), and minimum (front row) flow levels measured at 22 monitoring
stations. The far left corner corresponds to the New Mexico border and
the right to the Gulf of Mexico. The vertical scale correctly portrays in
perspective the relative flow peaks corresponding to the areas where
U.S. and Mexican tributaries enter the river. The horizontal axis is only
a rough approximation of relative horizontal distances along the length
of the Rio Grande/Río Bravo.

Figure 1.5

Reservoirs in the Rio Grande/Río Bravo Basin

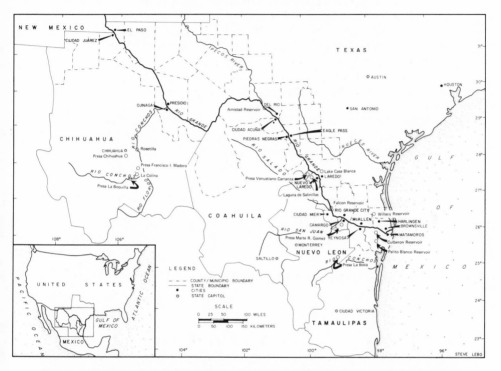

Source: International Boundary and Water Commission of the United States and Mexico--United States Section, *Flow of the Rio Grande and Related Data, 1980*, Water Bulletin 50 (El Paso: IBWC, 1980), pp. 83-87.

aquifer at a given point can affect the volume of water available throughout the aquifer.

This chapter has shown how surface water availability depends upon and to a large extent is defined by the climate of the region. However, given a climate and a potential water availability, man can manipulate available supplies to fit human needs. Dams are built to ensure year-round flow and to control flooding; diversions are constructed to move water to where it is most useful.

The only critically low flow areas along the river are between El Paso/Juárez and Presidio/Ojinaga. Reservoirs and water projects upstream from the New Mexico/Texas border and diversions at El Paso/Juárez limit streamflow in the region. Downstream from the confluence of the Río Conchos, streamflow is relatively abundant. Amistad and Falcon reservoirs control and make possible large diversions at Maverick and Anzalduas canals and lesser diversions in the far downstream reaches of the river. The Rio Grande/Río Bravo region exemplifies the human ability to manage water for multiple uses.

REFERENCES

1. Raymond Johnson, *The Rio Grande* (East Sussex: 1 Publishers, Inc., 1981), p. 49.

2. H. B. Mendieta, *Reconnaissance of the Chemical Quality of Surface Waters of the Rio Grande Basin, Texas*, Report 180 (Austin: Texas Water Development Board, 1974), p. 3.

3. John C. Day, *Managing the Lower Rio Grande*, Report 125 (Chicago: University of Chicago, Department of Geography, 1970), p. 7.

4. U.S. Department of State, International Boundary and Water Commission of the United States and Mexico--United States Section, *Flow of the Rio Grande and Related Data, 1978*, Water Bulletin 48 (El Paso: IBWC, 1978), pp. 159-160.

5. Mendieta, *Reconnaissance of the Chemical Quality*, p. 3.

6. Day, *Managing the Lower Rio Grande*, p. 7.

7. Neal E. Armstrong, "Anticipating Transboundary Water Needs and Issues in the Mexico-United States Border Region in the Rio Grande Basin," *Natural Resources Journal* 22, no. 4 (October 1982): 880.

8. Mendieta, *Reconnaissance of the Chemical Quality*, p. 3.

9. Day, *Managing the Lower Rio Grande*, p. 7.

10. Ibid., p. 12.

11. Felipe Ochoa y Asociados, S.C., *Estudio de la Calidad del Agua en la Cuence del Río Bravo* (Mexico, D.F.: Secretaría de Agricultura y Recursos Hidráulicos, September 1978), pp. 7-9 (Limited Distribution Document).

12. Johnson, *The Rio Grande*, p. 49.

13. Texas Water Commission, *Reconnaissance Investigation of the Ground-Water Resources of the Rio Grande Basin, Texas*, Bulletin 6502 (Austin: TWC, July 1965), p. U15.

14. Lyndon B. Johnson School of Public Affairs, Policy Research Project, "Water Quality in the Rio Grande River: An Environmental and Institutional Assessment," Austin, May 1982, p. 7 (Draft Report).

15. U.S. Department of State, International Boundary and Water Commission of the United States and Mexico--United States Section, *Flow of the Rio Grande and Related Data 1980*, Water Bulletin 50 (El Paso: IBWC, 1980), pp. 26, 28.

16. Interview with Mr. Jose E. Cisneros Prado, Director of the District for the Secretaría de Agricultura y Rescursos Hidráulicos, Ciudad Acuña, Mexico, March 14, 1983.

17. Comisión del Plan Nacional Hidráulico, *Esquema de Desarrollo Hidráulicos para la Cuenca del Río Bravo* (Mexico, D.F.: Secretaría de Agricultura y Recursos Hidráulicos, December 1980), pp. 12-15 (Internal Document).

18. Ibid., p. 13.

19. Ibid., pp. 16-19.

20. Comisión del Plan Nacional Hidráulico, *Esquema de Desarrollo Hidráulico*, p. 19.

21. Ibid., p. 23.

22. Ibid.

23. Ibid.

24. U.S. Department of State, *Flow of the Rio Grande*, Water Bulletin 50, pp. 9-78, pp. 83-87.

25. Texas Department of Water Resources, "Water for Texas: Planning for the Future," Austin, February 1983, p. III-23-8 (Draft).

26. Department of Economic and Social Affairs, *Ground-Water Storage and Artificial Recharge*, Natural Resources Water Series no. 2 (New York: United Nations, 1975), pp. 17 and 29.

27. Texas Water Development Board, "Continuing Water Resources Planning and Development for Texas: Phase I," Austin, May 1977, p. 751 (Draft Report).

28. Day, *Managing the Lower Rio Grande*, p. 11.

29. U.S. Department of State, *Flow of the Rio Grande*, Water Bulletin 50, p. 144.

30. Ibid.

31. Ibid.

32. Texas Water Development Board, "Continuing Water Resources Planning and Development," p. 751.

33. Ibid.

34. Armstrong, "Anticipating Transboundary Water Needs," p. 883.

35. U.S. Department of State, *Flow of the Rio Grande*, Water Bulletin 50, pp. 146-47.

36. Ibid.

37. Ibid., p. 148.

38. Ibid.

39. Ibid., pp. 144-45.

40. Ibid.

41. Mark A. Rugen, David A. Lewis, and Irwin J. Benedict, *Evapotranspiration as a Method of Disposing of Septic Tank Effluent*, Report G75-185 (San Antonio: Raba Associates, for the Edwards Underground Water District, 1977), p. 13.

42. Ibid., p. 16.

43. Armstrong, "Anticipating Transboundary Water Needs," p. 884.

44. Texas Water Development Board, *Proposed Water Resources Development in the Rio Grande Basin* (Austin: TWDB, 1966), p. 27.

45. Armstrong, "Anticipating Transboundary Water Needs," p. 844.

46. U.S. Department of State, *Flow of the Rio Grande*, Water Bulletin 50, pp. 83-87.

CHAPTER 2
Surface Water Quality

The use of water is related to its quality. As early as 2,400 years ago Hippocrates associated disease with poor water quality: "Whoever wishes to investigate medicine properly should . . . consider most attentively the waters which the inhabitants use . . . for cooking" (1).

Wildlife, agricultural, and even industrial use of water can be limited by low quality. For example, many freshwater fish cannot propagate at a pH level below 6.0 or live when pH falls below 4.0 (2). Fourteen of the major crop plants exhibit a 50 percent decrease in yield if irrigated with water with a high salts content (3). Water cannot be used for some industrial cooling where silica content exceeds 50 milligrams per liter (mg/l) (4). These examples illustrate a vital point: Water resource policy in a region must take account of the fitness of the water for beneficial and desired uses.

The purpose of this chapter is to consider quality issues for both surface water and groundwater along the border between Texas and Mexico. To do so, we shall consider the following:

● the nature of water quality standards and the institutions involved with implementing such standards.

● water quality monitoring activities used to assess compliance with the standards.

● the status of surface water and groundwater along the Texas and Mexico sides of the border.

● some restrictions on use that such water quality may indicate.

21

WATER QUALITY STANDARDS

Proper examination of water quality and its management along the border requires consideration of both the Mexican and American approaches to setting and administering standards. Water quality standards in the United States and Mexico are established in a comparable manner. Scientific evidence is gathered as a basis for criteria relating water contaminant effects on beneficial water uses. Water quality standards are then developed in order to restrict human disposal of pollutants to levels that will not degrade natural water below the quality required for each intended use.

The United States has a decentralized system for the development and administration of water quality standards based upon two federal acts and associated amendments. Pursuant to these laws, the federal Environmental Protection Agency (EPA) develops specific regulations. State and local governments have primacy in enforcement activities and in permitting procedures for drinking water systems and effluent dischargers. States may adopt federal criteria or set more stringent standards. Local governments may also set stricter standards for drinking water or pollutant emission sources.

A more centralized system operates in Mexico. One federal law and its amendments set the guidelines for water quality standards. General surface water requirements and effluent limits are established by one federal agency and drinking water standards by another. The standards are applied by the two federal agencies and by local governments, which also share responsibilities for compliance and enforcement activities.

United States

Surface water standards in the United States are based on the Federal Water Pollution Control Act of 1972 (P.L. 92-500) and its amendment, the Clean Water Act of 1977 (P.L. 95-217). Two goals were set out in the original Act: zero discharge of pollutants and "fishable-swimmable" national water. The law mandated:

● establishment by EPA of effluent limits for municipal and industrial dischargers.

● prohibition of discharges except as permitted through the National Pollution Discharge Elimination System (NPDES) administered either by EPA or by states with plans approved by EPA.

● development of state water quality management planning programs under Section 208 for the control of both nonpoint and point source pollution.

The Act required EPA to develop criteria relating water quality to various beneficial uses including (a) the protection and propagation of fish and wildlife, (b) recreation in and on water, and (c) water supply for domestic, industrial, and agricultural uses (5).

Regulations restricting contaminants in groundwater in the United States have been aimed at groundwater used as a drinking water supply. Although common carrier drinking water standards have been in effect since 1914, national limits on contaminants in community water supplies date from the federal Safe Drinking Water Act of 1974 (P.L. 93-523). The Act expanded EPA's authority over residential community water systems and called for the issuance of updated Interim Primary Drinking Water Regulations by the agency.

In principle, both P.L. 92-500 and P.L. 93-523 assign to the states the primary responsibility for setting, administering, and enforcing water quality standards. All state actions, however, are subject to EPA approval (6). In Texas, surface water quality responsibilities lie with the Texas Department of Water Resources (TDWR); the Texas Department of Health (TDH) is charged with assuring drinking water quality.

Surface water standards set by TDWR consist of three parts:

● general criteria applicable to most surface water.

● numerical criteria for chlorides, sulfates, total dissolved solids, dissolved oxygen, pH, fecal coliform bacteria, and temperature.

● criteria for water uses deemed desirable for specific surface waters. Uses recognized by TDWR are contact recreation, noncontact recreation, propagation of fish and wildlife, and domestic raw water supply (7).

In order to administer the water quality program, TDWR divides water into stream segments. The twelve segments included in the Río Grande/Río Bravo basin are listed and described in Table 2.1. Figure

2.1 indicates the location of each segment. TDWR determines the desirable uses for each segment and assigns numerical standards for the quality of the natural water. Table 2.2 indicates the water uses and numerical standards assigned to the Rio Grande/Río Bravo basin segments.

These standards are water quality goals, not enforceable upper bounds on contaminant levels. The TDWR uses water quality information to assess compliance with stream standards or to award permits to point source dischargers. The effluent limits required by these permits are legally binding.

Authority for the permit program for municipal and industrial dischargers is shared between EPA and TDWR, which has not been delegated NPDES authority. The TDWR now administers the program, which requires each discharger to obtain a permit specifying allowable discharges. Since standards relate water quality to water use, the allowable effluent limits vary for different municipalities, industries, and geographic areas. Dischargers are required to submit periodic water quality reports to TDWR, which in turn may take administrative or legal action against repeated violators. EPA may veto a state permit; all permit limitations must comply with EPA regulations (8).

In order to assess surface water quality conditions, TDWR employs a classification and ranking system for stream segments. Segments are categorized as either water quality limited or effluent limited. A segment is water quality limited if significant violations of water quality standards exist, if the effluent limits for point source dischargers are not stringent enough to ensure compliance with the standards, or if advanced municipal treatment is necessary to protect water quality in the area. An effluent-limited classification is assigned to all other segments that meet requirements for industrial and municipal dischargers. The segment ranking system takes into account numerical quality criteria, water use, and population density (9). Table 2.1 summarizes the 1980 classification and ranking of the Rio Grande/Río Bravo basin segments.

The adoption and administration of drinking water standards in Texas is the responsibility of TDH, although local governments may elect to adopt and administer more stringent standards. TDH issues permits to all public water systems according to two categories: community water system and noncommunity water system. A community water system serves at least 15 one-family residential unit connections year-round or regularly serves at least 25 residents. Noncommunity water systems include all other public water systems (10).

Drinking water standards vary within these two classifications. In general, TDH has adopted EPA drinking water standards. Texas standards for bacteriological, physical, organic chemical, and radioactive materials are identical to EPA regulations and are applied to all community water systems. Of these, the only EPA regulations that TDH

Figure 2.1

Location of TDWR Stream Segments

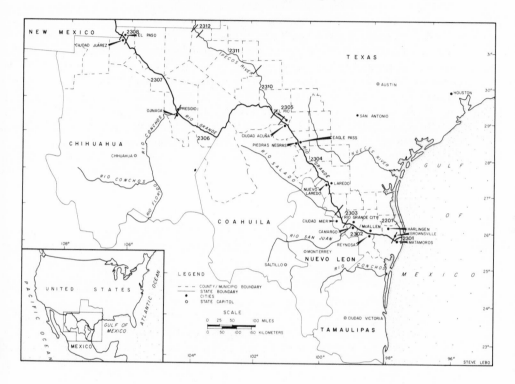

applies to noncommunity systems are bacteriological and physical contaminant limits. TDH standards for chemical substances differ slightly from EPA regulations (11). Table 2.3 details Texas's requirements according to system category.

In order to assess compliance with the standards, TDH requires periodic self-reporting from all permitted users. TDH issues public notices for violations of maximum contaminant levels or failures to perform water quality monitoring. Repeated violations can be grounds for administrative or legal action by TDH in cooperation with the State of Texas (12).

Although the Federal Water Pollution Control Act granted EPA and the states some authority over groundwater pollution, no standards or regulations aimed at the general quality of groundwater have been developed (13).

Mexico

The basis for surface water quality standards in Mexico is the Ley Federal para Prevenir y Controlar la Contaminación Ambiental (Federal Law for the Prevention and Control of Environmental Contamination), which was passed March 23, 1971. The goals of the legislation were to prevent:

● contamination of receiving water.

● interference with the purification process of water.

● modification, interference, or alteration in water usage.

● disturbances of the functioning of sewage water systems, the capacity of basins, or other national property.

The Secretaría de Agricultura y Recursos Hidráulicos (SARH; Secretary of Agricultural and Water Resources) was assigned responsibility for the development and administration of surface water quality standards in Mexico. The Act required actions by SARH to be taken in cooperation with the Secretaría de Salubridad y Asistencia (SSA; Secretary of Public Health), which also has primacy with respect to drinking water programs (14). In 1973, SARH issued the Reglamento para la Prevención y Control de la Contaminación de Aguas (Regulations for the Prevention and Control of the Contamination of Waters), which set standards for general surface water quality and for industrial and municipal discharges into surface water.

As in the United States, desired water uses are the basis of the water quality standards. Table 2.4 lists SARH's four water-use classifications for general stream quality. Maximum allowable contaminant levels for eleven parameters were established for each use by SARH; these are detailed in Tables 2.5 and 2.6. Although these water quality requirements have been adopted, they are not applicable to water in the Rio Grande/Río Bravo basin. SARH has yet to issue its determination of the desired water uses for the border region (15).

SARH regulates industrial and municipal dischargers by requiring a minimum level of internal treatment as a first phase for controlling contaminant levels. Table 2.7 outlines maximum levels of pollutant emissions for all dischargers. SARH regulations call for the implementation of five successive phases of pollution control for assuring efficient multiple use of water resources and maintaining regional environmental quality. In the second stage, industries are asked to add additional equipment to limit effluents. The third, fourth, and fifth phases call for the establishment of regional water quality control districts; industrial and municipal dischargers are asked to build or maintain shared treatment facilities (16). There is some indication that industrial dischargers are failing to comply with the phase-two requirements and there is little evidence of any progress toward regional waste treatment along the Mexican side of the Rio Grande/Río Bravo basin (17).

The law requires SARH, SSA, and state authorities to be jointly responsible for assessment of compliance with the phase-one standards. SARH can impose fines and seek imprisonment of regular violators. Mexican states may adopt more stringent standards. In practice, however, only SARH is active in the Texas/Mexico border region through its regional offices in Ciudad Acuña (Coahuila) and Río Bravo (Tamaulipas).

City governments are required to comply with the Mexican federal drinking water standards and are responsible for monitoring water quality (see Table 2.8). SSA can intervene or impose sanctions if a contaminated supply is found to cause continual health problems. Available information indicates little activity by either local governments or SSA with respect to compliance monitoring (18).

In summary, there are similarities and differences in the development and administration of water quality standards in the United States and Mexico. In both countries the laws establishing water quality goals are federal. However, the system of institutions involved with setting, administering, and enforcing the specific standards is more centralized in Mexico than in the United States. Table 2.9 summarizes these distinctions.

MONITORING

The way to assess whether water resources do in fact achieve quality standards is to test the concentrations of various parameters. Four separate agencies currently monitor water quality: TDWR, the U.S. Geological Survey (USGS), SARH, and the International Boundary and Water Commission. TDWR operates a statewide monitoring network that includes stations along the Rio Grande/Río Bravo, the Pecos River, and the Devils River. TDWR also conducts groundwater tests on a less regular basis. USGS conducts groundwater quality surveys on request and also operates water quality monitoring stations along the Rio Grande/Río Bravo, the Pecos River, and the Devils River. SARH operates a network of regional monitoring laboratories in Mexico (19). IBWC operates stations along the Rio Grande/Río Bravo and its tributaries along both the U.S. and the Mexican side. Figure 2.2 illustrates the location of all the monitoring stations located along the Rio Grande and its major tributaries.

IBWC was established in 1889 to mediate boundary and water conflicts along the Mexico/United States border. IBWC's first water monitoring station was established in 1889 at El Paso to gauge streamflows. Sampling for suspended silt and total salts was initiated in the 1920s and regular chemical contaminant sampling began in 1930 at the El Paso station. Additional stations were later added above and below Falcon Dam (built in 1954) and above and below Amistad Dam (built in 1969). By 1980, IBWC operated a network of twenty stations along the river to measure various parameters on a monthly basis. In addition, IBWC evaluates surface water quality at a station on each of the following tributaries: Río Conchos, Río Salado, Río San Juan, Devils River, Pecos River, along with a station at Morillo Drain (20).

The USGS, which was established in 1883, began sporadic monitoring of the Rio Grande/Río Bravo in the mid-1930s and regular testing of organic chemical quality in the mid-1950s. In response to the Federal Water Pollution Control Act of 1972 (P.L. 92-500), USGS began operating the National Stream Quality Accounting Network (NASQAN) in 1974. NASQAN includes five stations along the Rio Grande/Río Bravo that evaluate surface water for organic, chemical, and pesticide parameters on a monthly basis. By 1978, most stations in

Figure 2.2

Location of Monitoring Stations along the Rio Grande/Río Bravo

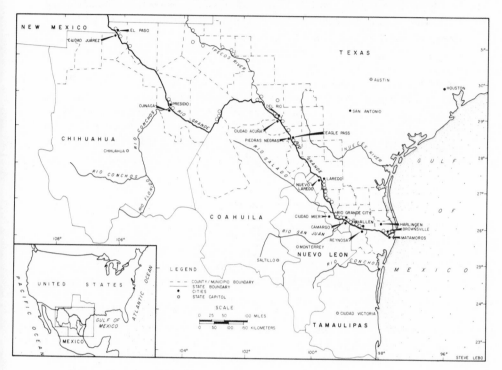

the basin included organic and biological contaminant testing in monitoring routines. In addition to its fixed station monitoring network, USGS conducts special intensive surveys of water quality on a contractual basis. According to USGS, possible federal budget reductions may force the closing of one or more stations along the Rio Grande/Rio Bravo within the next few years (21).

Water quality monitoring by the State of Texas dates back to 1957, when the State Department of Health (now Texas Department of Health) and the State Game and Fish Commission (now Texas Parks and Wildlife Department) initiated a surface water sampling program. After the 1967 creation of the Texas Water Quality Board, a statewide monitoring network was established to succeed the joint Health Department/Parks and Wildlife program. After the 1977 consolidation of the Texas Water Rights Commission, the Texas Water Development Board, and the Texas Water Quality Board into the Texas Department of Water Resources, TDWR Enforcement and Field Operations Division took over control of the statewide monitoring network. TDWR 1980 records indicated monitoring stations at thirty-five sites along the Rio Grande/Rio Bravo; since that time TDWR has closed or changed several of its monitoring stations (22).

Both TDWR and USGS regularly assess groundwater quality. Over the past few years TDWR has built and maintained a statewide network of monitoring wells. A specific well is designated as a monitoring site for a geographic area on the basis of whether there is historical information, accurate knowledge of the aquifer, and a good rapport with the well owner or operator. TDWR attempts to test each area every five years for general inorganic chemical quality. TDWR does not test organic quality or pesticides because agency policy views these parameters as health-related. If there is reason to believe that there has been some deterioration in the quality of water in a particular area, TDWR may undertake a more detailed analysis or refer the problem to TDH. USGS does extensive testing and reporting in response to specific requests but does not operate a regular monitoring system for groundwater quality (23).

In Mexico SARH oversees a national network of Water Quality Control Offices and a regional network of monitoring laboratories. The Rio Grande/Rio Bravo is located within two of these regions (see Figure 2.3). Groundwater quality monitoring is not done on a regular basis, although mobile laboratories can investigate groundwater problems as they arise. Since results of surface water and groundwater testing are generally not published, they are difficult to obtain (24).

Historical information on water quality has been published periodically by IBWC, USGS, and TDWR. Current information on quality for both surface and groundwater is available in computer output form from the Texas Natural Resources Information System (TNRIS), which

Figure 2.3

Location of SARH Regional Monitoring Laboratories

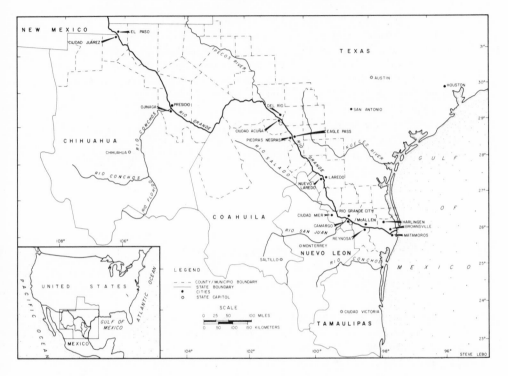

also stores self-reporting data for industrial and municipal dischargers. Surface water quality stations have primarily monitored chemical quality; biological or health-related information is sparse prior to the mid-1970s.

Several factors make it difficult to analyze accurately any historical changes in water quality: the numerous agencies involved in water quality testing, the different parameters each monitors, and the dearth of regular data from all but a few stations over time. Reports for many years include limited information. Chemical quality measures, primarily salinity, are the only information consistently available.

Despite these difficulties it is still possible to develop a general picture of surface water and groundwater quality in the basin from the historical records. It is also possible to identify those areas where water quality problems do exist, have existed in the past, or may exist in the future.

SURFACE WATER QUALITY

Evaluating water quality requires water analysis in four interrelated dimensions: time, site, source, and contaminant. It is necessary to consider both the fluctuations of contaminant levels over time at a given site, and the behavior of contaminants over space at a given time. A complete evaluation of water conditions in a region requires the consideration of both surface water and groundwater if both occur in the area, as there may be a physical interrelationship between them. For example, contaminated surface water may eventually degrade regional groundwater supplies.

Certain distinctions between surface water and groundwater affect use. For example, as groundwater moves more slowly than surface water, the time and space factors may be viewed differently. Contaminated groundwater may not be diluted as quickly as polluted surface water and may therefore pose a more difficult problem. Also, monitoring practices for groundwater and surface water differ. Surface water quality is examined more regularly and for more contaminants than groundwater. Although the study of groundwater and surface water should not be separate in principle, in practice inadequate information on their interrelationships restricts the reliability of conclusions about surface water and groundwater interactions.

Although agencies monitor numerous parameters along the Rio Grande/Río Bravo, this section will focus on the few that reflect general levels of water quality. These parameters include those for which TDWR sets standards: dissolved oxygen (DO), total dissolved solids (TDS), fecal coliform, pH, temperature, sulfates, and chlorides.

TDWR classifies only two segments in the basin as water quality limited, its indication of contamination. The other portions of surface water along the Rio Grande/Río Bravo are considered adequate for all intended uses. Salinity has been and continues to be the major water quality problem in the basin. High levels of TDS, attributable to natural causes in the arid region, have been reported over time (25).

TDWR has determined that violations occur in segment 2308 at El Paso and in segment 2201, the Arroyo Colorado. These relate respectively to a sewage treatment plant and to municipal effluents and toxic substances from unexplained sources (26). Although water quality may be adequate along most of the region, this assertion should not obscure the fact that site-specific problems do exist and could get worse. The following subsections review water quality patterns in the upper, middle, and lower reaches to assess parameter levels that may indicate current or potential problems.

Upper Rio Grande/Río Bravo Reach

The upper Rio Grande/Río Bravo reach stretches approximately 484 river miles (774 km) from the Texas/New Mexico border near El Paso to the Amistad Reservoir and includes TDWR segments 2308, 2307, and 2306. Below El Paso/Juárez, the river is virtually dry because of irrigation diversions to Mexico. Perennial flow does not resume until the Río Conchos contributes its water 284 miles (454 km) downstream from El Paso.

The flow of the Rio Grande/Río Bravo as it enters Texas is derived principally from snowmelt and is regulated by the Elephant Butte and Caballo reservoirs in New Mexico. The discharge weighted average concentration of TDS has been near 500 mg/l; samples have generally not exceeded 1,000 mg/l in this water (27). Water quality in the upper Rio Grande/Río Bravo reach is characterized by generally high levels of TDS, especially in the El Paso/Juárez and Fort Quitman areas. Other problems in the region include high fecal coliform concentrations near El Paso/Juárez and at the confluence of the Río Conchos, and pesticide residues near the Río Conchos (28).

Segment 2308 runs from the Texas/New Mexico border to the Riverside Diversion Dam in El Paso. Water quality in the area varies by season. TDWR has classified this segment as water quality limited, owing to the relatively high concentrations of TDS, fecal coliform, sulfates, and chlorides. Fecal coliform levels have consistently exceeded the upper criteria limits. From 1975 to 1979, TDWR reported between 15 and 80 percent of samples in violation (29). TDS concentrations in the segment are generally higher than those upstream, owing to irrigation return flows. During the spring and summer, TDS concentrations are similar to upstream values as this is the period during which water is released from the reservoirs. In the fall and winter when no water is released, TDS levels are higher, ranging between 200 and 2,078 mg/l over the four year period 1979 through 1982 (30).

Segment 2307 includes the river flow from the Riverside Diversion Dam in El Paso to the confluence with the Río Conchos near Presidio. The main water quality monitoring stations in this segment are at Fort Quitman and above Presidio. As most of the water passing El Paso is diverted for irrigation and municipal use in Mexico, little flow occurs in this area. During an extended period between 1951 and 1959, for example, flows came only from local storm runoff and streams below the irrigated valley upstream. When waters do flow past the irrigated areas before reaching Fort Quitman, TDS concentrations in the past have been high (31). The wide fluctuations in TDS levels in 1955, a drought year, indicate the salinity problem in the area (see Figure 2.4). IBWC 1980 reports found TDS levels to range between 1,530 and 5,650 mg/l (32).

Water that passes the station at Fort Quitman often does not reach Presidio, as it may be diverted through evapotranspiration, diversion, and seepage. Since most of the water at upper Presidio is contributed by mountain creeks or arroyos below Fort Quitman, TDS concentrations are reduced below levels upstream. In 1980, TDS levels ranged between 748 and 3,260 mg/l (33).

Below Fort Quitman, three-quarters of the water in the Rio Grande/Río Bravo is contributed by Mexican tributaries. The Río Conchos flows into the main channel just below Presidio, marking the beginning of segment 2306 which stretches over 312 miles (499 km) to the Amistad Reservoir. High TDS, elevated fecal coliform levels, and irregular pesticide residues characterize the segment.

TDS levels decrease as the Río Conchos joins the Rio Grande/Río Bravo. TDS levels in the Río Conchos ranged between 803 and 1,200 mg/l in 1980 (34). Unfortunately there is little information available on water quality in the Rio Grande/Río Bravo just below the Río Conchos confluence. IBWC has operated a monitoring station there for many years but has only measured specific conductance. By the time the water

(continued on page 36)

Figure 2.4

TDS Concentrations along the Rio Grande/Río Bravo

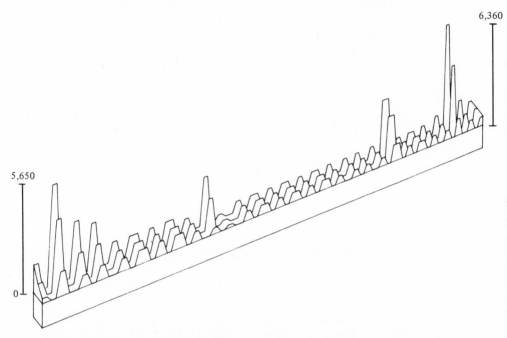

Source: International Boundary and Water Commission of the United States and Mexico--United States Section, *Flow of the Rio Grande and Related Data: 1980*, Water Bulletin Number 50 (El Paso: IBWC, 1980), pp. 88-113.

The figure illustrates 1980 total dissolved solid levels in the main stem of the river, ranging from less than 200 mg/l to more than 6,000 mg/l. The three rows represent the maximum (back row), median (middle row), and minimum (front row) TDS levels measured at 29 monitoring stations. The far left corner corresponds to the New Mexico border and the far right to the Gulf of Mexico. The vertical scale correctly portrays the four relative TDS maxima--close to Fort Quitman, Langtry, the Río Salado, and the Morillo Drain. The horizontal axis is only a rough approximation of relative horizontal distances along the length of the Rio Grande/Río Bravo.

reaches Foster Ranch at Langtry, Texas, TDS concentrations fluctuate less, ranging in 1980 from 664 to 889 mg/l (35).

A study compiled for SARH in 1974 reported excessive fecal coliform and sulfate concentrations in the Río Conchos. Total coliform levels were found to be as high as 24,000 colonies/100 ml; the median was 13,200 colonies/100 ml (36). TDWR has reported excessive fecal coliform levels below the Río Conchos confluence and has also reported residues of DDT and endrin in the area (37). Levels of these contaminants fall as the river reaches Foster Ranch. Fecal coliform concentrations generally were below 500 colonies/100 ml in the years 1979 through 1982, although excessive levels have been occasionally reported (38). No pesticides were found in monitoring tests at Foster Ranch in 1980 (39).

Middle Rio Grande/Río Bravo Reach

The middle region of the Rio Grande/Río Bravo stretches 1,014 miles (1,622 km) and flows from the Amistad Reservoir to the Falcon International Reservoir; this region includes TDWR segments 2305, 2304, 2312, 2311, 2310, and 2309. Two tributaries, the Pecos and Devils rivers, drain into the main channel from the Texas side in this reach. Water quality in and below the Amistad Reservoir is generally good, although high fecal coliform levels have been reported in the Del Río/Ciudad Acuña and Laredo/Nuevo Laredo areas. The Pecos River extends from the Red Bluff Reservoir to Langtry, Texas, where it joins the Rio Grande/Río Bravo. Water from the Pecos is highly saline; excessive fecal coliform levels have occasionally been reported in the Red Bluff Reservoir. Oil spills and oil seepage in the Pecos are major water quality problems. The Devils River generally exhibits good quality.

Segment 2305, the Amistad Reservoir, stretches 74.7 miles (119.5 km) and includes the sites where the Pecos and Devils rivers flow into the reservoir from the Texas side. Water quality in the reservoir is generally better than the segment criteria limits. High TDS levels have been reported occasionally in the reservoir (40).

Segment 2304 runs 226.5 miles (362.4 km) from the Amistad Dam to Falcon Lake. TDS levels recorded at the monitoring station 2.2 miles (3.5 km) downstream of the dam contained between 602 and 673 mg/l of TDS (41). Monitoring stations at Laredo/Nuevo Laredo indicated good chemical quality in 1980 as well. Fecal coliform levels, however, were reported as high as 61,000 colonies/100 ml in 1980 (42). Below Laredo/Nuevo Laredo, fecal coliform concentrations as high as 280,000 colonies/100 ml were reported (43).

Segment 2312 is the Red Bluff Reservoir which impounds the Pecos River. This segment extends 11.4 miles (18 km) from the Texas/New Mexico state line to the Red Bluff Dam. In the past the water in the reservoir has been characterized as saline. In the period from 1978 through 1982, TDS levels ranged from 913 to 14,096 mg/l; the mean level for this period was 4,524 mg/l. These high TDS levels are primarily attributable to natural causes. Although fecal coliform and dissolved oxygen measures fall within the standards, several fish kills have occurred in the reservoir in recent years (44).

Segment 2311 runs from Red Bluff Dam to near Pandale, Texas. In the past TDS levels in this segment have been high. In 1980, TDS concentrations ranged from 6,170 to 16,200 mg/l with a mean level of 9,195 mg/l (45). These high TDS levels are attributable to salt deposits in the drainage basin, irrigation return flows, and inflow of brine from oil-field activity. Oil-field practices and the presence of oil have greatly affected water quality in this segment. Fish kills have resulted from oil spills in the area (46). A 1974 TDWR study reported oil seepage into the Pecos near Iraan, Texas, and attributed it to both natural and man-made causes. Seepage occurs as groundwater moves through oil-bearing deposits, flushing out some oil which is discharged through springs and builds up in sloughs. Following heavy rains and the resulting rise in the river, this oil flows into the Pecos. Because of the expense and complications involved in intercepting the oil-water discharge, one recent TDWR study recommended no corrective action (48).

Segment 2310 stretches 49 miles (78.4 km) along the Pecos, from near Pandale, Texas, to the Amistad Reservoir. The only apparent water quality concern in this segment is salinity. In 1980, USGS reported TDS levels ranging from 382 to 2,900 mg/l, reduced from those upstream (48).

Segment 2309, the Devils River, extends 136.6 miles (218.6 km) northwest from the Amistad Reservoir. Water quality in the Devils River is usually excellent.

Lower Rio Grande/Río Bravo Reach

The lower Rio Grande/Río Bravo reach stretches 348.2 miles (557 km) from the Falcon International Reservoir to the Gulf of Mexico and includes TDWR segments 2303, 2302, 2301, and 2201. Two tributaries, the Río Salado and the Río San Juan, contribute to the main stream from the Mexican side in this segment. Although water quality in the Falcon Reservoir is generally good, much of the water in the segment is highly saline and registers excessive fecal coliform levels. Four drains contribute irrigation return flows with high TDS concentrations near Fort Ringgold, Texas. Industrial and municipal effluents also affect river quality. Pesticide residues have been reported in the Brownsville/Matamoros area. Segment 2201, the Arroyo Colorado, is classified by TDWR as water quality limited owing to fecal coliform and pesticide contamination.

Segment 2303 is the Falcon Reservoir, which stretches 68.3 miles (109 km) from the confluence of the Río Salado to Falcon Dam and impounds the Rio Grande/Río Bravo. Water quality in the reservoir is generally good.

Segment 2302 runs 230.6 miles (369 km) from Falcon Dam to the International Bridge at Brownsville. Much of the segment is character-ized by high TDS and fecal coliform levels, as well as by contamination from industrial effluents. Both releases from Falcon Dam and irrigation practices affect TDS concentrations in the upper portion of the segment. The narrowness of the drainage basin in this area prevents agricultural return flows from being a major consideration. Farther south, however, irrigation return flows enter the river from four drains, two above and two below Fort Ringgold. Water from these drains is considerably more saline than that in the river (49). Early efforts to prevent saline con-tamination in the Rio Grande/Río Bravo resulted in the construction of a 75-mile (120 km) diversion channel from a fifth drain, the Morillo, to the Gulf of Mexico.

Contaminant measurements of the river in proximity to the mu-nicipal areas indicate industrial and municipal contamination. At Nueva Guerro, for example, untreated municipal waste-water is dis-charged directly into the Rio Grande/Río Bravo. Water quality meas-urements in this area have reported excessive levels of suspended silt, greases and oils, phosphates, and sodium--possibly a reflection of in-dustrial wastes (50). Farther south, at the confluence with the Río San Juan, high sediment, salinity, phosphorous, and greases and oils have also been reported (51). Water quality measurements at discharge points in Río Bravo, Reynosa, and at Anzalduas Dam have detected excessive silt, chlorides, and oils (52).

Segment 2301 includes 49.3 river miles (78.9 km), stretching from the International Bridge at Brownsville to the Gulf of Mexico. Oxygen depletion, high fecal coliform concentrations, and reported pesticides contribute to the water quality problems in this segment. During the period 1978 through 1982, fecal coliform measures ranged from zero to 20,000 colonies/100 ml (53). A 1981 study by the consulting firm of Black and Veach reported concentrations of pesticides in the area, both in irrigation return flows and in municipal drinking water supplied from the Rio Grande/Río Bravo (54).

Segment 2201, the Arroyo Colorado, stretches 80.9 miles (129.4 km) from Hidalgo County to where it enters the Laguna Madre between Willacy and Cameron counties. This segment is classified as water quality limited by TDWR. Dissolved oxygen levels are periodically below segment standards. Fecal coliform and TDS levels are frequently excessive. During the period 1978 through 1982, TDS concentrations ranged from 700 to 16,000 mg/l (55). Hazardous levels of pesticides such as dieldrin, DDD, DDE, and PCBs have been reported in recent years (56). The TDWR's list of additional problems in the segment includes excessive organic nitrogen, COD, volatile residues, chlordane, dissolved cadmium, iron, nickel, mercury, and manganese (57). Municipal effluent and irrigation return flows contribute to many of these problems.

WATER QUALITY AND WATER USE

Although the Mexican and Texas authorities generally view quality of the surface water of the Rio Grande/Río Bravo as acceptable, there clearly are some water quality concerns both throughout the basin and in specific areas. Salinity is the overriding water quality problem in both present and historical times. Figure 2.4 illustrates the TDS concentration along the river in 1980. Excessive concentrations of fecal coliform are present in the more populated areas of the region (see Figure 2.5). Water near some municipal areas also exhibits pollution resulting from industrial effluents. The presence of pesticides, while not a basinwide concern, may be a problem in the upper and lower reaches of the area (see Table 2.10 for a list of water quality problem areas). At excessive levels, these contaminants can restrict the four major water uses in the region: public water supply, livestock watering, irrigation, and industry.

(continued on page 41)

Figure 2.5

Fecal Coliform Levels along the Rio Grande/Río Bravo

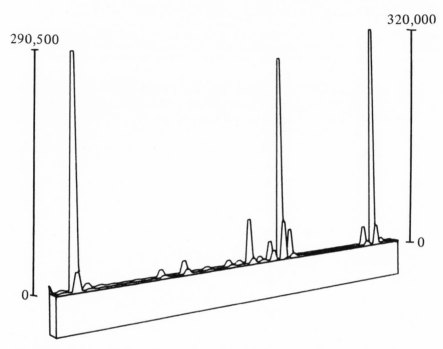

Source: International Boundary and Water Commission of the United States and Mexico--United States Section, *Flow of the Rio Grande and Related Data: 1980*, Water Bulletin Number 50 (El Paso: IBWC, 1980), pp. 88-113.

The figure illustrates 1980 fecal coliform levels in the main stem of the river, ranging from approximately zero to more than 300,000 colonies/100 ml. The three rows represent the maximum (back row), median (middle row), and minimum (front row) fecal coliform levels measured at 34 monitoring stations. The far left corner corresponds to the New Mexico border and the far right to the Gulf of Mexico. The vertical scale correctly portrays in perspective the relative coliform peaks below El Paso and Juarez, Laredo and Nuevo Laredo, and Brownsville and Matamoros. The horizontal axis is only a rough approximation of relative horizontal distances along the length of the Rio Grande/Río Bravo.

For example, the U.S. Environmental Protection Agency has determined that water containing high concentrations of TDS is not considered acceptable for human consumption because of possible physiological effects and unfavorable taste (58). Water containing under 3,000 mg/l of soluble salts is most suitable for livestock watering; levels beyond this amount may result in physiological damage, animal death, and economic losses (59). Plants irrigated with saline water may fail or experience reduced yields, although the degree to which plants are affected depends upon many factors including the level of salinity, soil type, and the method of irrigation (60). Salinity may also affect industrial water use by contributing to corrosion and increasing production costs.

Water exhibiting high fecal coliform levels is usually unsuitable for public water supply and irrigation. The likelihood of a health hazard increases as the geometric mean of fecal coliform densities exceeds 2,000 colonies/100 ml in raw water. Water exhibiting fecal coliform densities above 1,000 colonies/100 ml may adversely affect human health when used to irrigate crops for human consumption (61).

The presence of pesticides in water is also a matter of concern. Both the extent of pesticide contamination in the Rio Grande/Río Bravo region and the effects of these substances in general are a matter of controversy. However, the occurrence of pesticides in raw water used for livestock watering, public water supplies, and irrigation may have harmful effects.

These few examples illustrate the point that water of poor quality can affect water use. However, the relationship between water quality and use is reciprocal; not only does water quality affect use, but the use of water may also have an impact on its quality. For example, under some circumstances, repeated irrigation use of saline water will increase salinity. Increasing population can overload a community waste-water treatment system, thereby causing water contamination. Industrial use of water without proper treatment may lead to pollutant discharge. Water resource management requires assessing and maintaining the capability of water to meet current and future needs. It must also consider the effects on water quality resulting from expanded or altered water use.

REFERENCES

1. Hippocrates, *On Airs, Waters, and Places*, quoted in Brian MacMahon and Thomas F. Pugh, *Epidemiology Principles and Practices* (Boston: Little, Brown and Co.,1970), p. 5.

2. Environmental Studies Board, *Water Quality Criteria 1972* (Washington, D.C.: U.S. Government Printing Office, 1973), p. 141.

3. Ibid., p. 325.

4. Ibid., p. 377.

5. *Federal Water Pollution Control Act of 1972*, (P.L. 92-500), sec. 304(a) and sec.101(a)(2) as cited in U.S. Environmental Protection Agency, *Quality Criteria for Water* (Washington, D.C.: U.S. Government Printing Office, 1977), p. ix.

6. Texas Department of Water Resources, *Surface Water Quality Standards* (Austin: TDWR, April 1981), p. 2; Texas Department of Health, Division of Water Hygiene, "Drinking Water Standards Governing Drinking Water Quality and Reporting Requirements for Public Water Supply Systems" (Austin: TDH, November 29, 1980), p. 1.

7. Texas Department of Water Resources, *Surface Water Quality Standards*, pp. 9-11.

8. Environmental Protection Agency, Office of Public Awareness, "A Guide to the Clean Water Act", Report No. OPA 129/8 (Washington, D.C.: U.S. Environmental Protection Agency, November 1978), pp. 13-14.

9. Texas Department of Water Resources, *The State of Texas Water Quality Inventory*, 6th ed., Report No. LP-59 (Austin: TDWR, 1982), pp. 3-5.

10. Texas Department of Health, Division of Water Hygiene, "Drinking Water Standards," p. 1.

11. Ibid., pp. 2, 10-12, 17.

12. Texas Department of Health, *Biennial Report: September 1, 1979-August 31, 1981* (Austin: TDH, 1981), pp. 127-128.

13. Council on Environmental Quality, *Environmental Quality-1980: The Eleventh Annual Report of the Council on Environmental Quality*, reprinted by the Library of Congress, Congressional Research Service, December 27, 1981, p. 97.

14. Secretaría de Agricultura y Recursos Hidráulicos, *Reglamento Para la Prevención y Control de la Contaminación de Aguas* (Mexico, D.F.: SARH, March 29, 1973), Chapter I, Article 2.

15. Interview with Dr. Pedro Martinez, Visiting Professor, Lyndon B. Johnson School of Public Affairs, The University of Texas, Austin, February 1983.

16. Secretaría de Agricultura y Recursos Hidráulicos, *Reglamento Para la Prevención y Control de la Contaminación de Aguas* (Mexico, D.F.: SARH, March 29, 1973), Chapter II, Article 16.

17. Interview with Dr. Pedro Martinez.

18. Ibid.

19. Ibid.

20. U.S. Section, International Boundary and Water Commission, *Flow of the Rio Grande and Related Data: 1980*, Water Bulletin No. 50 (El Paso: IBWC, U.S. Section, 1981), pp. 92, 96, 97, 102, 105.

21. Interview with Helen Davidson, staff member, U.S. Geological Survey, Water Resources Division, Austin, November 1982.

22. Texas Department of Water Resources, Enforcement and Field Operations Division, "Monitoring Stations along the Rio Grande as of 1982," Austin, 1982 (Unpublished Computer Printouts).

23. Interview with Gerald Baum, Data and Engineering Services Division, Texas Department of Water Resources, Austin, November 1982.

24. Interview with Dr. Pedro Martinez, Visiting Professor, Lyndon B. Johnson School of Public Affairs, The University of Texas, Austin, February 1983.

25. H. B. Mendieta, *Reconnaissance of the Chemical Quality of Surface Waters of the Rio Grande Basin, Texas*, Report No. 180 (Austin: Texas Water Development Board, March 1974), pp. 12-26.

26. Texas Department of Water Resources, *The State of Texas Water Quality Inventory: 1980*, 5th ed. (Austin: TDWR, 1981), p. 436, 447.

27. H. B. Mendieta, *Reconnaissance of Chemical Quality*, pp. 17-18.

28. Texas Department of Water Resources, *The State of Texas Water Quality Inventory: 1980*, p. 446.

29. Ibid., p. 455.

30. Texas Department of Water Resources, *The State of Texas Water Quality Inventory: 1982*, 6th ed., p. 449.

31. H. B. Mendieta, *Reconnaissance of Chemical Quality*, p. 18.

32. International Boundary and Water Commission, *Flow of the Rio Grande*, p. 90.

33. Ibid., p. 91.

34. Ibid., p. 92.

35. U. S. Geological Survey, *Water Resources Data for Texas: Water Year 1980*, vol. 3 (Austin: U.S. Geological Survey, 1981), p. 491.

36. Felipe Ochoa y Asociados, *Estudio de la Calidad del Agua en la Cuenca del Río Bravo* (Mexico, D.F.: Secretaría de Agricultura y Recursos Hidráulicos, September 1978), p. 68 (Limited Distribution Document).

37. Texas Department of Water Resources, *Texas Water Quality Inventory*, p. 447.

38. Ibid.

39. United States Geological Survey, *Water Resources Data*, pp. 492-493.

40. Texas Department of Water Resources, *Texas Water Quality Inventory*, p. 445.

41. International Boundary and Water Commission, *Flow of the Rio Grande*, p. 98.

42. Ibid., p. 100.

43. United States Geological Survey, *Water Resources Data for Texas: Water Year 1980*, p. 541.

44. Texas Department of Water Resources, *Water Quality Inventory*, p. 453.

45. U. S. Geological Survey, *Water Resources Data for Texas*, p. 507.

46. Texas Department of Water Resources, *Water Quality Inventory*, p. 452.

47. Charles Blackwell, *Oil Seepage into the Pecos River* (Austin: Texas Water Quality Board, 1974), p. 7.

48. U. S. Geological Survey, *Water Resources Data for Texas*, p. 525.

49. Steve Warshaw, *Water Quality Segment Report for Segment No. 2302, Rio Grande* (Austin: Texas Water Quality Board, 1974), p. 24.

50. Felipe Ochoa y Asociados, *Estudio de la Calidad*, p. 91.

51. Ibid., pp. 56-57.

52. Ibid., pp. 58-62.

53. Texas Department of Water Resources, *Texas Water Quality Inventory*, p. 441.

54. Black and Veach, *Toxic Substance Sampling Program* (Dallas: Lower Rio Grande Development Council, June 1981), pp. I-1, I-2.

55. Texas Department of Water Resources, *Texas Water Quality Inventory*, p. 430.

56. Ibid., p. 429.

57. Ibid.

58. Environmental Studies Board, *Water Quality Criteria 1972*, p. 90.

59. Ibid., p. 308.

60. Ibid., p. 58.

61. Ibid., p. 351.

CHAPTER 3
Groundwater

Mexicans and Texans who reside along the banks of the Rio Grande/Río Bravo share the same groundwater resource, but they do not share similar water use patterns. Groundwater resources are currently the most important water sources for the domestic, municipal, industrial, agricultural, power, and mining consumers of water in several communities. Other areas barely utilize groundwater at all, but heavily compete or even overuse the Rio Grande/Río Bravo.

This chapter deals with both the availability and quality of groundwater along the Rio Grande/Río Bravo. It also discusses areas of potential conflicts among water users. Two of the points of greatest future conflict are likely to be the valley along the lower Rio Grande/Río Bravo and the El Paso/Ciudad Juárez area. Although several aquifers occur in the lower Rio Grande/Río Bravo Valley, use is currently limited by contaminants occurring at levels that affect water users. Groundwater in the El Paso/Ciudad Juárez area is being depleted as water is withdrawn at a rate higher than the water recharge.

GROUNDWATER AVAILABILITY

An aquifer is any geologic formation that contains sufficient saturated permeable material to yield significant quantities of water to wells and springs. A major aquifer is defined as one which yields large quantities of water over a comparatively large area. A minor aquifer is one which yields enough water to supply a small area (1).

Table 3.1 lists the major and minor aquifers in the Rio Grande/ Río Bravo basin. Figure 3.1 shows the locations of major aquifers in the basin. Major aquifers occur all along the Mexico/Texas border; the largest is the Edwards-Trinity (Plateau) Aquifer, which occurs in the middle reach.

The next subsections discuss water availability in each of the major aquifers. Table 3.2 contains estimates of water storage capacity and recharge of these major aquifers. Table 3.3 lists the projected average annual groundwater availability in these aquifers.

Mesilla and Hueco Bolsons

The Mesilla and Hueco bolsons are the major groundwater resources of the El Paso/Ciudad Juárez area. Ciudad Juárez/El Paso residents currently withdraw groundwater at the highest rate anywhere in the Rio Grande/Río Bravo basin.

The Mesilla Bolson Aquifer lies west of the Franklin Mountains and extends over a wide area of State of Chihuahua in Mexico and states of New Mexico and Texas in the United States. The basin of the Mesilla Bolson is composed of clay, silt, sand, and some gravel (2). In this area water needs have in the past exceeded the volume of surface water flow in the Rio Grande/Río Bravo. Hence the Rio Grande Alluvium, which overlies the bolson deposits in the Mesilla and El Paso valleys, is an important source of shallow groundwater for supplemental irrigation. For example, over 57 thousand acre-feet (70 million cubic meters) of water were withdrawn from the Mesilla and Hueco bolsons in 1975, 27 thousand acre-feet (33.5 million cubic meters) for municipal and industrial purposes and 30 thousand acre-feet (37 million cubic meters) to irrigate agricultural areas.

Figure 3.1

Texas Rio Grande Aquifers

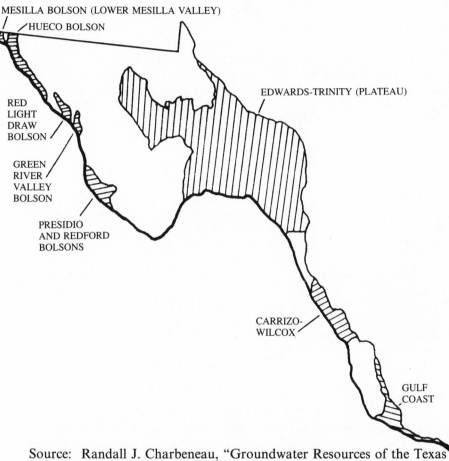

Source: Randall J. Charbeneau, "Groundwater Resources of the Texas Rio Grande Basin," *Natural Resources Journal* 22, no. 4 (October 1982): 958. Reprinted with permission from *Natural Resources Journal* , published by the University of New Mexico School of Law, Albuquerque, NM 87131.

The Hueco Bolson Aquifer includes areas in Chihuahua, Mexico, and Texas and New Mexico in the United States. The northern part of this bolson lies between the Organ and Franklin mountains on the east. The southern part straddles several mountain ridges in Chihuahua, Mexico, on the west, and the Diablo Plateau and Finley, Malone, and Quitman mountains on the east.

The bolson deposits have a total thickness that ranges from two thousand feet (610 meters) near El Paso/Ciudad Juárez area in the Mesilla Bolson, to 9,000 feet (2,740 meters) in the Hueco Bolson area. In some places, these deposits contain fresh groundwater to depths of about 1,200 feet (370 meters). Most of the wells in the Rio Grande Alluvium are used for irrigation. Well yields range from 32 to 3,012 gallons per minute (2 to 190 liters per second) (3).

Water-level maps suggest that water moves from north to south, parallel to the Rio Grande/Río Bravo. The lower Mesilla Valley serves as a discharge zone, as the narrow outlet at El Paso del Norte restricts groundwater outflow from the valley, forcing most of the water discharged to evapotranspirate. These maps do not indicate that significant volumes of groundwater move south across the border (4). Table 3.4 shows the volume of groundwater withdrawn from the lower Mesilla Valley Bolson.

Water levels in the shallow portion of the Mesilla Bolson Aquifer have not changed much over time, as the aquifer is replenished by infiltration from canals, the Rio Grande/Río Bravo, and return flows from agricultural lands. Water levels in the medium and deep aquifers declined during pumping from 1957 to 1960. Pumping in this area is currently balanced by groundwater inflow, mainly from the shallow aquifer. The water-level decline in the medium aquifer is 15 to 20 feet/4.5 to 6.0 meters, while the decline in the deep aquifer is 30 to 70 feet/9 to 21 meters (5).

Water recharge to the Hueco Bolson occurs along the bordering mountains and also from the Rio Grande/Río Bravo. The annual recharge around the northern edge of the bolson, including areas in New Mexico and Chihuahua, is about 6 thousand acre-feet (7.4 million cubic meters) of water per year (6). Heavy pumping near El Paso/Ciudad Juárez has drawn down water levels by about 95 feet (30 meters) in the downtown area (7). Table 3.5 shows the groundwater stored in the western portion of the Rio Grande/Río Bravo basin.

Table 3.6 shows an estimate of El Paso's future water requirements to the year 2030. These requirements include water for domestic, municipal, manufacturing, power, mining, and livestock uses; water for irrigation is not included. Column three lists the projected surface water deliveries from the Rio Grande/Río Bravo to El Paso. The difference between the total projected requirements and the surface water deliveries is made up by groundwater withdrawals. This water would be supplied from the Canutillo well fields in the lower Mesilla Valley and from the Mesa and Montana well fields in the Hueco Bolson (8).

One analyst has estimated that by 1989 the pumping of the Mesilla Bolson Aquifer will reach a maximum sustainable yield; more rapid extraction could deplete the aquifer by 2030 or permit saline water intrusion. Columns 5 and 6 of Table 3.6 list the portion of the groundwater requirements which cannot be met by the Mesilla Bolson; such demand may be met with withdrawals from the Hueco Bolson Aquifer. One estimate is that by 2030 about 25 percent of the recoverable fresh groundwater storage will remain in these bolsons (9).

If the water of these bolsons is depleted or degraded, users in El Paso County have several alternate water sources: import other fresh water, treat saline water, pump less groundwater, or reclaim existing return flows. The closest possible Texas sources of fresh water in limited quantities are (a) the Red Light Draw Bolson located 100 miles (160 kilometers) southeast of El Paso, and (b) the southern portion of the Salt Bolson, 150 miles (240 kilometers) southeast of the city. The nearest sources of fresh groundwater for import are in New Mexico. One analyst estimated that in 1974 the extension of the Hueco Bolson into New Mexico stored as much as 6.2 million acre-feet (7,650 million cubic meters) of fresh water and as much as 2.9 million acre-feet (3,580 million cubic meters) of slightly saline water (10).

Red Light Draw Bolson

The Red Light Draw Bolson lies between the Quitman Mountains to the West and the Eagle and Indio mountains to the east. This bolson's water is currently withdrawn for livestock wells. Water-level contours show that recharge moves from the bordering mountains to the axis of the valley and then southeastward toward the Rio Grande/Río Bravo (11).

One estimate is that about 600,000 acre-feet (740 million cubic meters) of fresh groundwater are stored in this aquifer. No problems are known to be associated with developing groundwater in this bolson. Not all of the water can be withdrawn because removal of fresh water may draw in adjacent saline water. The annual recharge is estimated to be about 2000 acre-feet (25 million cubic meters). The fresh groundwater levels up in the valley are 200 to 400 feet (60 to 120 meters) higher than the water levels in areas of saline water along the Rio Grande/Río Bravo (12).

Green River Valley, Presidio and Redford Bolsons

The Green River Valley Bolson lies between the Indio and the Van Horn mountains along the Green River, an ephemeral stream which is a tributary of the Rio Grande/Río Bravo. The Presidio Bolson extends along the Rio Grande/Río Bravo from Candelaria, Texas, to outcrops of volcanic rock 6 to 10 miles (10 to 16 kilometers) southeast of Presidio, Texas. The Redford Bolson continues along the Rio Grande/Río Bravo for another 12 miles (19 kilometers) southeast of the volcanic outcrops (13). Currently this water is used for livestock, irrigation, and domestic supply.

Table 3.7 lists the storage and recharge volumes of the Green River Valley and Presidio and Redford bolsons. As the bolsons' fresh water is at a higher elevation than the poorer quality water along the Rio Grande/Río Bravo, no problems are anticipated with extracting fresh groundwater, unless the water levels are lowered excessively (14).

Edwards Trinity (Plateau) Aquifer

The Edwards-Trinity (Plateau) Aquifer underlies the Edwards Plateau and extends westward into the eastern part of the Trans-Pecos region of Texas to include Val Verde, Terrell, and part of Brewster counties. It is composed of water-saturated sand and sandstone of the Trinity Group and limestone of the overlying Fredericksburg and Washita Groups of Cretaceous age. The thickness of the limestone reaches 1,000 feet (300 meters), while the sand is usually less than 100 feet (30 meters) (15). The measured spring flows of the Rio Grande/Río Bravo,

Pecos, and Devils rivers suggest that the annual recharge of this aquifer is approximately 514 thousand acre-feet (634 million cubic meters) (16).

Carriso-Wilcox Aquifer

The Carrizo-Wilcox Aquifer is located in the lower part of the Río Grande/Río Bravo basin. It is composed of an upper unit, the Carrizo Formation of the Claiborne Group, and a lower unit, the Wilcox Group (17). Along its northern and western edge, the aquifer is exposed at the surface; it is recharged by precipitation and streams crossing the outcrop area (18). The saturated thickness of the aquifer ranges from 200 to 700 feet (60 to 215 meters) (19).

One current estimate of the recoverable storage of the Carrizo-Wilcox Aquifer is 160 thousand acre-feet (197 million cubic meters) of water. This aquifer is recharged by 13,700 acre-feet (16.9 million cubic meters) of water annually (20). Groundwater levels have declined in several heavily pumped irrigation areas located over this aquifer, such as the Winter Garden area and the municipal/industrial well fields north of Lufkin, Texas. In 1964, more than 360 thousand acre-feet (444 million cubic meters) of water were pumped from the aquifer for irrigation purposes in the Winter Garden area, an amount greatly exceeding the estimated annual natural recharge (21).

Gulf Coast Aquifer

The Gulf Coast Aquifer in the lowest part of the basin is composed of the Catahoula, Oakville, Lagarto, Goliad, Lissie, and Beaumont formations. These surface deposits of recent age consist of interbedded sand, gravel, silt, and clay. The maximum thickness of this fresh to slightly saline water is approximately 500 feet (150 meters) (22).

The annual recharge of the Gulf Coast Aquifer is estimated to be 11,400 acre-feet (14 million cubic meters) of water; it is recharged by precipitation on the surface and seepage from streams crossing the area (23). Even though the rate of natural recharge is sufficient to sustain present pumpage rates, sustained heavy pumpage has caused serious problems in local areas. There has been land surface subsidence, increased chloride content in the groundwater of the southwest portion of the aquifer, and salt-water encroachment along the coast (24).

Aquifers Located in Mexico

The annual recharge of groundwater aquifers located on the Mexican side of the Rio Grande/Río Bravo basin is of the order of 1.5 to 1.7 million acre-feet (1,812 million to 1,972 million cubic meters) (see Table 3.8). The total storage volume of water available other than in Nuevo León is 11.3 million acre-feet (13,960 million cubic meters). One study has estimated that in Nuevo León, an additional 851,245 acre-feet (1,050 million cubic meters) of water may be available for each meter of depression of the static level of the aquifer (25).

GROUNDWATER QUALITY

The factors that influence groundwater quality range from the mineral composition of the subsurface environment to contaminants added by human activities. Some factors affect the concentration and forms of pollutants in all groundwater, including the source of the water, the mineral composition of the rocks through which the water moves, and the rate of groundwater movement. Precipitation that percolates to the water table contains dissolved carbon dioxide from the air; both react with organic matter in the soil and rock particles, dissolving them and forming new compounds.

Groundwater can be polluted if contaminants are discharged into an aquifer recharge area, through wells drilled in the aquifer, or into surface streams that feed it. The sources of pollution can be more numerous and more widely distributed than surface water contamination. Groundwater can store contamination; hence, groundwater treatment can be difficult, costly, or impractical. Effluent control and enforcement techniques used for surface water may be irrelevant because the fate of the pollutants cannot be traced. Monitoring is more difficult and expensive for groundwater than for surface water, because wells have to be dug. If wells are scattered arbitrarily over the aquifer, many localized plumes of contamination can be missed (26).

Mesilla and Hueco Bolsons

The lower Mesilla Valley is an area of complex subsurface sedimentation, hydrology, and soil characteristics. This complexity affects both lateral and vertical groundwater quality. The sediments in this area vary from clays to coarse gravels, which can affect groundwater quality by restricting water movement. Where horizontal movement is restricted, a buildup in dissolved solids may occur because there is no lateral influx of fresh water. Where vertical movement is contained, zones of saline water may occur above or below the zones of better water quality (27). Water quality in the shallow aquifer can be influenced by surface water quality of the Rio Grande/Río Bravo because of water infiltration. In general, the water in the shallow aquifer is of poorer quality than the deeper groundwater. Table 3.9 shows the source, significance, and concentration range of dissolved mineral constituents of the water of the Rio Grande Alluvium and Hueco and Mesilla bolsons. The groundwater north of Canutillo, Texas, contains from 260 to 2,300 mg/l of total dissolved solids (TDS); it is classified as slightly saline. In the south area, the groundwater is very saline, with TDS concentrations as high as 24,800 mg/l. Water from the deep aquifer in the Canutillo, Texas, and Anthony, New Mexico, area commonly contains less than 300 mg/l of TDS (28).

Southward from El Paso the water in the aquifer increases in mineral content. Shallow aquifer water is generally of better quality than the water in the underlying bolson deposits. This groundwater is used for irrigation and to leach the soil of accumulated salt (29). The city of Fabens, Texas, obtains its public water supply from bolson deposits of about 300 feet (100 meters) depth. The groundwater contains 520 mg/l of TDS and has a hardness of 129 mg/l (moderately hard). An El Paso test well (located three miles or 4.8 kilometers south of Fabens) at a depth of 1,783 feet (543 meters) found groundwater that contained 180 mg/l of chloride, 476 mg/l of sulfate, and 3.4 mg/l of fluoride (30). The fluoride and sulfate concentrations are well in excess of the drinking water contaminant limits recommended by the U.S. Environmental Protection Agency.

The largest volume of fresh water with TDS concentrations less than 500 mg/l resides in the Hueco Bolson, adjacent to the Franklin Mountains. This body of water occurs in sediments that include coarse debris eroded from the mountains and alluvium deposited by the ancestral Rio Grande/Río Bravo. Recharge occurs mainly from runoff infiltration along the Franklin and Organ mountains. This combination of both a source of recharge and the presence of coarse grained/permeable sediments results in a relatively rapid and deep circulation of fresh water (31).

Water in the bolson deposits beneath the Rio Grande/Río Bravo alluvium, southeast of Ysleta, Texas, is moderately saline, with TDS concentrations greater than 5,000 mg/l (32). The water in the southern part of the Hueco Bolson, north of Fort Hancock, Texas, is slightly saline (1,000 to 3,000 mg/l of TDS) but hard (greater than 180 mg/l of TDS). Sulfate and fluoride concentrations have been measured as 500 mg/l and 5.0 mg/l, respectively (33).

Red Light Draw and Green River Valley Bolsons

Groundwater in most of the Red Light Draw and Green River Valley bolsons is fresh, containing less than 500 mg/l of TDS (34). Groundwater in the alluvium along the Rio Grande/Río Bravo at the lower end of these bolsons is of poorer quality, ranging from slightly saline (1,000 to 3,000 mg/l of TDS) to very saline (3,000 to 10,000 mg/l of TDS) (35). In the Indian Hot Springs area west of lower Red Light Draw Bolson, part of the groundwater is hot, with temperatures ranging between 104° to 122°F (40° and 50°C) (36).

Presidio and Redford Bolsons

Most of the groundwater in the Presidio and Redford bolsons is fresh. Northwest of Cibolo Creek, in the area between the foothills and floodplain, the bolson fill is mostly clay and silt. The groundwater in the alluvium along the stream channel is fresh, and water in the fine-grained bolson fill is moderately saline. In the Cibolo/Alamito Creek area, the entire section of bolson fill is coarser grained and may contain fresh water (37).

Northwest of Presidio, Texas, along the Rio Grande/Río Bravo, the groundwater quality ranges from fresh to very saline; most is moderately saline (38). Below Presidio, groundwater in the alluvium is slightly saline. After the Río Conchos enters the Rio Grande/Río Bravo, 4 miles (6 kilometers) above Presidio, it becomes a perennial stream. This inflow from the Río Conchos and Alamito Creek has maintained the quality of the groundwater along the Rio Grande/Río Bravo. Groundwater quality varies both with area and time, reflecting periodic recharge from the river. Poor quality of the groundwater aquifers in this area results from the discharge of groundwater by evapotranspiration along the floodplain of the river (39).

Wells that furnish water to the cities of Candelaria and Presidio, Texas, yield water that has low concentrations of TDS, sulfates, and chlorides. However, the fluoride concentration in this water exceeds the maximum contaminant level established by EPA for drinking water (40).

Edwards-Trinity (Plateau) Aquifer

Only limited water supplies are withdrawn from the Edwards-Trinity (Plateau) Aquifer for use along the Rio Grande/Río Bravo. Part or all of the water supply of the cities of Fort Stockton, McCamey, Iran, Sheffield, and Sanderson, Texas, comes from this aquifer. With the exception of Fort Stockton these potable water supplies meet the standards established by the Texas Department of Health. The chloride and sulfate concentrations in the Fort Stockton water supply are each about

500 mg/l, high enough to impart a salty taste and a laxative effect to the water. Water of better quality is available west of this city, where groundwater contains less than 1,000 mg/l TDS and the chloride and sulfate concentrations are less than 200 mg/l (41).

A limited volume of water is pumped from the Edwards-Trinity (Plateau) Aquifer to irrigate agriculture in Reeves and Pecos counties. Much of the water contains more than 2,000 mg/l of TDS. Figure 3.2 shows the major irrigation areas located in this area. Repeated infiltration of irrigation water, heavy applications of fertilizers, and perching of water in Reeves County have saturated the soils with minerals, reducing crop yields. The TDS concentration in this area ranges from 1,000 to more than 3,000 mg/l (42).

The major irrigation areas in Pecos County include North Coyanosa, South Coyanosa, Fort Stockton-León-Belding, Girvin, and Bakersfield. In the North Coyanosa area, groundwater quality is suitable for irrigation and industrial purposes. The TDS and fluoride concentrations here are higher than the upper limits accepted by the Texas Department of Health for public water supplies (43). Table 3.10 lists the TDS concentration of the groundwater in this area.

Carrizo-Wilcox Aquifer

Through most of its Texas extensions, the Carrizo-Wilcox Aquifer yields fresh to slightly saline water (44). However, saline water does occur in areas such as the Winter Garden area, where geological formations overlying the aquifer contain saline water. Owing to improper water well completion and failure of well casings from corrosion, the artesian heads of the aquifer have declined. There may also have been an influx of saline water (45). Excessive pumpage in certain areas is causing reversals of the hydraulic gradient in the aquifer. The changes may result in encroachment of water of poorer quality into regions having better-quality water (46).

Figure 3.2

Irrigation Areas in the Edwards-Trinity (Plateau) Aquifer

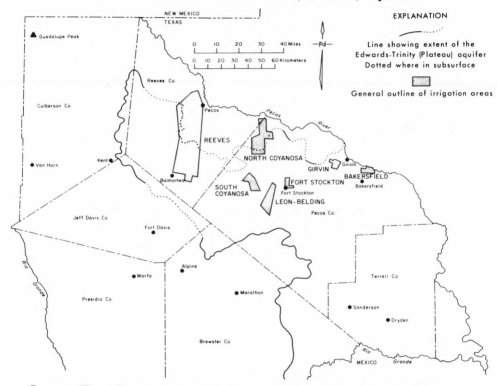

Source: Texas Department of Water Resources, *Occurrence and Quality of Groundwater in the Edwards-Trinity (Plateau) Aquifer in the Trans-Pecos Region of Texas*, Report 255 (Austin: Texas Department of Water Resources, September 1980), p. 15.

Gulf Coast Aquifer

Groundwater throughout most of the eastern part of the Gulf Coast Aquifer is low in TDS (less than 500 mg/l). Sodium and bicarbonate are principal contaminants, and the water is comparatively soft (47).

In the southern part of the aquifer, groundwater is more saline. In some areas highly saline water overlies or underlies the fresh water (48). The TDS concentration in this part of the aquifer is between 1,000 and 1,500 mg/l. However, there are areas in Aransas, Calhoun, Cameron, Hidalgo, Kenedy, Kleberg, Nueces, San Patricio, and Willacy counties where no appreciable amounts of fresh to slightly saline groundwater can be found (49).

On gulf shore islands, groundwater suitable for domestic and livestock requirements may be found in shallow sands. Portions of this water are unacceptable for prolonged irrigation because of high salinity, alkalinity, or both problems (50).

Aquifers Located in Mexico

Although the groundwater quality of aquifers on the Mexican side of the Rio Grande/Río Bravo is believed to be generally good, local problems may occur. In the lowest regions of the bolsons, such as north of Aldama and in the Juárez Valley, Chihuahua, the groundwater contains high concentrations of TDS. Some of the aquifers occurring under calcium carbonate rock are polluted by surface disposal of solids and industrial wastes. In the Bajo Bravo and Bajo San Juan rivers, the high volume of withdrawn groundwater may be provoking an intrusion of saline water from clay layers overlying the aquifer (51).

CONCLUSIONS

The heaviest groundwater users in the United States are the states which border Mexico. The water laws and institutions in these states do not completely control groundwater withdrawals, despite the fact that these resources must be shared by the two countries.

The challenges of managing groundwater in Texas include (a) virtual absence of law for resolving conflicts among pumpers, (b) lack of coordination of groundwater and surface water rights, (c) absence of programs to determine and assure optimum rates of depletion of non-recharging aquifers, and (d) absence of programs to protect recharging aquifers from damage due to excessive withdrawals. However, the prevailing view of Texans today appears to be that, although the state does face serious water problems that could get worse, no major changes in groundwater law or institutions are needed at present.

Although Texas has virtually no control over groundwater resources, the Mexican federal government has strong legal controls. The national government, through the *Secretaría de Agricultura y Recursos Hidráulicos* (SARH) (Ministry of Agriculture and Water Resources), has established the pumpage rate for each aquifer as well as the water price to be charged to each user.

The significant population and economic growth projected in the Rio Grande/Río Bravo basin complicates efforts to manage groundwater. As all surface water has been "allocated," any growth in water consumption means increasing groundwater withdrawals.

Although several studies have been made in relation to the groundwater resources along the Rio Grande/Río Bravo, it is difficult to make a quantitative assessment of the availability and quality of the water of the aquifers in this area. Many studies on Texas aquifers are broad and lack specific information; this may stem from the fact that groundwater in Texas is an uncontrolled private good. Occasionally information contained in one report is even contradicted by data presented in another study. Project participants were unable to find more than a few Mexican groundwater studies.

Despite these uncertainties, some general conclusions regarding groundwater management along the Texas/Mexico border can be drawn. The metropolitan area of Ciudad Juárez/El Paso, with a population of over one million inhabitants, depends on the Hueco Bolson Aquifer for the municipal and part of the agricultural water supply. Both cities currently pump at a rate faster than the groundwater in this aquifer is being recharged. Water supplies for this area will become more distant, more expensive, more scarce, and of lower quality due to lateral and vertical encroachment of saline water from adjacent saline water sands. Problems similar to those of the Ciudad Juárez/El Paso area may occur in the future in other metropolitan areas along the Rio Grande/Río Bravo, where cities on both sides of the border share the same groundwater reservoir.

Some analysts have proposed allocating jurisdiction over shared Mexican/United States groundwater resources to the International Boundary and Water Commission (IBWC). Regardless of what institutional arrangements the two nations may adopt, comprehensive studies should be undertaken to assess the present availability and quality of groundwater resources.

REFERENCES

1. Texas Department of Water Resources, *Groundwater Availability in Texas,* Report 238 (Austin: TDWR, September 1979), p. 4.

2. Randall J. Charbeneau, "Groundwater Resources of the Texas Río Grande Basin," *Natural Resources Journal* 22, no. 4 (October 1982):957.

3. Texas Department of Water Resources, *Groundwater Availability,* p. 25.

4. Charbeneau, p. 957.

5. Texas Department of Water Resources, *Groundwater Availability,* p. 25.

6. Texas Department of Water Resources, *Availability of Fresh and Slightly Saline Ground Water in the Basins of Westernmost Texas,* Report 256 (Austin: TDWR, September 1980), p. 93.

7. Texas Department of Water Resources, *Groundwater Availability,* p. 25.

8. Ibid.

9. Texas Department of Water Resources, *Availability,* p. 93.

10. Ibid, p. 97.

11. Charbeneau, p. 957.

12. Texas Department of Water Resources, *Groundwater Availability,* p. 33.

13. Texas Department of Water Resources, *Availability,* p. 71.

14. Charbeneau, p. 957.

15. Texas Department of Water Resources, *Availability,* p. 45.

16. Charbeneau, p. 957.

17. Ibid.

18. Texas Water Development Board, *The Texas Water Plan* (Austin: Texas Department of Water Resources, January 1982), p. II-9.

19. Charbeneau, p. 957.

20. Ibid.

21. Texas Water Development Board, *The Texas Water Plan,* p. II-9.

22. Charbeneau, p. 957.

23. Texas Water Development Board, *The Texas Water Plan,* p. II-10.

24. Texas Department of Water Resources, *Groundwater Availability,* p. 39.

25. Secretaría de Agricultura y Recursos Hidráulicos, *Esquema de Desarrollo Hidráulico para la Cuenca del Río Bravo* (Mexico, D.F.: SARH, December 1980), p. 34 (Internal Document).

26. A. Ludwick Teclaff, "Principles of Transboundary Groundwater Pollution Control," *Natural Resources Journal* 22, no. 4 (October 1982): 1065.

27. Texas Department of Water Resources, *Ground Water Development in the El Paso Region, Texas with Emphasis on the Lower El Paso Valley,* Report 246 (Austin: TDWR, June 1980), p. 39.

28. Ibid.

29. Texas Water Commission, *Reconnaissance Investigations of the Ground-Water Resources of the Rio Grande Basin, Texas,* Bulletin 6502, Second printing (Austin: TWC, August 1973), p. U-54.

30. Ibid.

31. Texas Department of Water Resources, *Groundwater Availability,* p. 25.

32. Ibid.

33. Texas Water Commission, *Reconnaissance Investigations,* p. U-54.

34. Texas Department of Water Resources, *Groundwater Availability,* p. 27.

35. Texas Department of Water Resources, *Availability,* p. 77.

36. Charbeneau, p. 957.

37. Texas Deparment of Water Resources, *Availability,* p. 83.

38. Texas Department of Water Resources, *Groundwater Availability,* p. 34.

39. Texas Department of Water Resources, *Availability,* p. 83.

40. Ibid.

41. Texas Department of Water Resources, *Occurrence and Quality of Groundwater in the Edwards-Trinity (Plateau) Aquifer in the Trans-Pecos Region of Texas,* Report 255 (Austin: TDWR, September 1980), p. 15.

42. Ibid.

43. Ibid.

44. Texas Department of Water Resources, *The State of Texas Water Quality Inventory,* 6th ed. (Austin: TDWR, 1982), p. 445.

45. Texas Water Development Board, *The Texas Water Plan,* p. II-9.

46. Texas Department of Water Resources, *Groundwater Availability,* p. 17.

47. Texas Water Development Board. *The Texas Water Plan,* p. II-10.

48. Ibid.

49. Texas Department of Water Resources, *Groundwater Availability,* p. 52.

50. Ibid.

51. Secretaría de Agricultura y Recursos Hidráulicos, *Esquema de Desarrollo,* p. 34.

CHAPTER 4

A Population Profile
of the Texas-Mexico
Border Region

Despite the fact that the Rio Grande/Río Bravo basin is relatively arid, the resident population is large and increasing. About 2.7 million persons lived in the region in 1980. One estimate forecasts a near doubling of the border-region population by the year 2000, and other projections foresee even more rapid growth.

This chapter presents a demographic profile of the Río Grande/Río Bravo region, north and south of the border. The first section describes the past and present population and its distribution within the cities, counties, and states covered by this report. Population densities as well as median ages, household sizes, and vital statistics are outlined. A second section discusses methodologies used to project future populations of the Texas/Mexico border regions along the Rio Grande/Río Bravo. The third section considers undocumented migration from Mexico to Texas and implications of population growth for water use along the border.

CURRENT POPULATION STATUS OF THE BORDER REGION

The Texas/Mexico border along the Rio Grande/Río Bravo is a region of geographic, climatic, and demographic contrast. In the northwestern corner of the border area are El Paso, Texas, and Ciudad Juárez, Chihuahua, partner communities with a combined population of greater than one million inhabitants (1). Twelve hundred miles (1,920 kilometers) downstream, at the southeastern terminus of the Rio Grande/Río Bravo, lie the sister cities of Brownsville and McAllen, Texas, and Reynosa and Matamoros, Tamaulipas, where approximately one million people live (2,3).

This border region, encompassing four Mexican states and sixteen Texas counties (see Figure 4.1), includes 248,249 square miles (642,965 square kilometers), or an area nearly the size of Texas. About 16 percent of the land area is in Texas and 84 percent in Mexico. The total population residing in this comprehensive region is greater than nine million inhabitants; about 13 percent of this population lives in Texas and 87 percent in Mexico. The population of the actual basin, however, is approximately 2.7 million; 1.4 million live in Mexican border cities and 1.2 million reside in Texas border counties. Although the Texas counties represent some of the poorest in the United States, the Mexico border region is paradoxically one of the most affluent areas in Mexico (4).

In many respects the border region acts as a whole, rather than two disparate parts. The area is linked by a common heritage of culture, language, and economics. As one analyst put the matter, "Whether industry locates in Brownsville or Matamoros does not matter. We are governed more by economic dictates. We are one economic unit" (5).

Texas

Among 254 Texas counties, thirteen share an extended border with Mexico: Cameron, El Paso, Hidalgo, Hudspeth, Kinney, Maverick, Presidio, Starr, Terrell, Val Verde, Webb, and Zapata. The southwestern boundary of two other counties, Dimmit and Culberson, are within ten miles of the Rio Grande/Río Bravo. Jeff Davis, another west Texas county, touches the northern edge of the state of Chihuahua, Mexico.

Four of these sixteen counties, Cameron, Hidalgo, Webb, and El Paso, are Standard Metropolitan Statistical Areas (SMSAs) that contain medium to large pockets of population. Nearly 1.1 million inhabitants live in these four areas. Approximately 140 thousand persons live in the remaining twelve counties. In 1980 seven of these counties had populations of less than 10,000; six contained between 10,000 and 100,000 inhabitants; and three counties (Cameron, El Paso, and Hidalgo) had a population exceeding 100,000 residents.

Figure 4.1

The Texas/Mexico Border Region

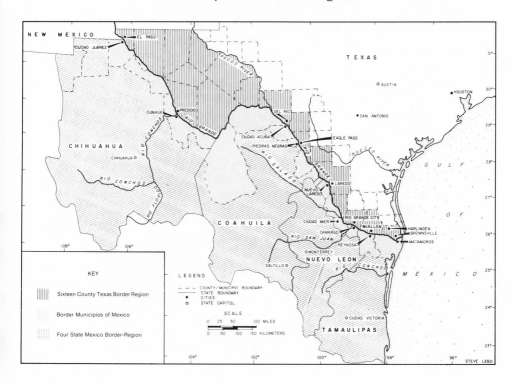

Table 4.1 shows how each county's population has changed over the past half-century. Although the Rio Grande/Río Bravo region grew as a whole, almost all of the seven counties with populations of less than 10,000 inhabitants have decreased in size. On the other hand, the SMSA counties have doubled or tripled in size over the past five decades. The Texas county with the greatest fifty-year continuous growth rate was Maverick County, which grew by 3.27 percent annually (6). Altogether Texas's Rio Grande/Río Bravo counties nearly tripled in population in the past fifty years. Table 4.2 compares these population growth rates. While Texas has been outstripping U.S. growth rates in recent years, the Rio Grande/Río Bravo counties have grown even faster.

Despite such growth, the areas along the Rio Grande/Río Bravo still have a relatively low population density. Table 4.3 lists population densities for the border counties. The sixteen counties comprise 39,500 square miles (102,305 square kilometers); an average of 30.6 persons live in each square mile (79.2 persons per square kilometer). This is about 58 percent of the population density of Texas (53.2 inhabitants per square mile or 137.8 inhabitants per square kilometer) and less than half of the 1980 United States population density (62.6 persons per square mile or 162.1 persons per square kilometer).

The population of the Texas counties is composed of 48.5 percent males and 51.5 percent females; these numbers may appear counter-intuitive. A common view is that males should be in the majority in this region due to its proximity to the border and the predominant "maleness" of immigrant agricultural laborers. The typical sex ratio (number of males divided by the number of female members of the population) is less than one in the United States, as there are more females than males. Predominantly "male" counties are the exception rather than the rule. Sex components of the population for each of the sixteen counties are presented in Table 4.4.

The data in Table 4.4 indicate that the median age is higher in counties that have experienced low or negative population growth. The implication is that as younger members of the population move away in search of greater economic opportunity, the median age shifts upward. The pattern of families with young children migrating out of the community also tends to increase the median age.

Another characteristic of the population in the study area is the large number of persons of Spanish origin. About 900 thousand, or 73.2 percent of the region's total population, fall in this category. Over 90 percent of the inhabitants of Maverick, Starr, and Webb counties are of Spanish origin. This rate far exceeds the rest of Texas and the United States, where the fraction of inhabitants of Spanish origin is 21 percent and 6.4 percent, respectively (7,8).

A particularly interesting characteristic of the border area indicated by the data in Table 4.5 is the large average household size of border families. In the United States the average household size in 1980 was 2.75 (9). Fourteen of the sixteen border counties exceed this mean. The average household size for the entire region is 3.57, almost 30 percent greater than the U.S. average and 26 percent greater than the Texas mean of 2.82 persons per household (10).

Although crude birth rates, death rates, and projected population changes in one year are only approximate measures of the direction and magnitude of growth within the region, one can infer from these statistics (see Table 4.6) that:

● The counties that have relatively high mean household sizes (for example, Hidalgo, Maverick, and Webb) have high rates of natural increase.

● The high rates of natural increase in nearly all the border counties account, in many cases, for less than half of the population increase.

In other words, this region's growth is characterized both by significant levels of family formation and, more importantly, by in-migration.

Mexico

Information on the population of the Mexican side of the border is somewhat less ample than Texas's census data. As of March 1982, the *X Censo de Población y Vivienda* (1980) represents the major source of data for Mexico. The forthcoming final 1980 census tabulations are likely to provide information not ordinarily available within the preliminary edition, such as the populations of each border municipio.

The population south of the Texas/Mexico border has grown even more rapidly than the nearby Texas counties (see Table 4.7). In twenty years (1960-80), the population of the border states adjacent to Texas has almost doubled in size. Nuevo León, which shares only a small portion of the boundary with Texas, has grown by 128 percent. The 1980 estimates indicate that the four state regions of Chihuahua, Coahuila, Nuevo León, and Tamaulipas (illustrated in Figure 4.1) have a combined population of 7.9 million people and are growing at a continuous annual rate of 3.10 percent.

The population of this four-state region has a greater proportion of male inhabitants than its counterpart region across the border. As indicated in Table 4.8, the sex ratio for the Mexican side of the border is 0.988, versus corresponding ratios of 0.941 and 0.968 for the Texas border area and for all of Texas, respectively. In comparison, the sex ratios for Mexico and the United States are 0.977 and 0.948. This rough measure suggests that the Mexican border contains a slightly greater proportion of male occupants than Mexico as a whole. Conversely, the Texas border region contains somewhat fewer male residents, compared with Texas or the United States. These differences are small in both cases, and it may be unreasonable to draw inferences from this evidence.

The number of residents in Mexico's border states may give an incorrect impression of the number of persons residing near the border, as the surface areas of the four states is over five times greater than Texas's sixteen county region. This area of 207,279 square miles (536,853 square kilometers) has a population density of 38.0 inhabitants per square mile (98.4 inhabitants per square kilometer), low relative to Mexico's average population density of 89.5 persons per square mile (231.8 persons per square kilometer) (see Table 4.9).

The border city populations are a better measure of the inhabitants residing along the Rio Grande/Río Bravo than the population of the four Mexican states (see Table 4.10). The urbanized population adjacent to the border is much smaller than the total number of people of the four states. It is therefore important to distinguish between the large population residing in the comprehensive region and the urban populations which punctuate the border. As additional information from the Mexican census becomes available, it may be possible to evaluate the pattern of growth in the Mexican border region more accurately.

BORDER REGION POPULATION PROJECTIONS FOR TEXAS

Population *projections*, as distinguished from predictions or forecasts, are estimates of the total size or composition of the population at a future date, given a set of assumptions. Projections are expectations of the future which hold only in so far as the assumptions are fulfilled.

The Texas Department of Health and the Texas Department of Water Resources project future populations for all Texas counties for use in statewide planning. Differences in the estimates of future inhabitants can affect the tax structure, bonding capacity, and the ability of a community to plan for government services. Furthermore, population projections may reflect governmental or private interests dependent on population growth for continued well-being. Careful evaluation of the assumptions and methods used in projection is important in any analysis of long-term population patterns.

Population projections for Mexico are prepared by the U.S. Bureau of the Census and the Mexican *Secretarío de Programación y Presupuesto*. As both organizations prepare estimates for all of Mexico and do not specifically project state population size, their methodologies will not be presented below. Extrapolations of state and city populations will be provided instead.

There are several distinct projection methodologies, three of which are discussed briefly below. These methods are extrapolation processes, demographic adjustment models, and econometric techniques.

Extrapolation

Extrapolation applies linear or exponential curve-fitting techniques to observations of the number of people over a period of time. If linear growth is assumed, extrapolation will project that a certain number of people enter the population each year. Linear regression methods are often applied to the data set in order to "best fit" a line to a set of observations (by minimizing the sum of squared errors between a set of estimates and the corresponding set of real observations). The resultant slope and intercept coefficients of this line allow an analyst to extrapolate subsequent years' population.

An exponential extrapolation assumes that the population grows by a certain percentage (as opposed to a certain number) each year. The data may be transformed into logarithmic form and then fitted by a line. The associated slope and intercept of the line equation then enable the projection of future populations.

Demographic Methods

The demographic approach decomposes population into components of births, deaths, and net migration. Each of these components has an associated probability. These probabilities are then applied to the population of the previous year in order to project future year populations. The probability of a certain cohort surviving until they "graduate" into the following cohort is called the survival ratio. The summation of survivors in all cohorts over time determines the expected number of persons at some future time.

Net migration (NM) is the difference between immigration (I) and emigration (E). Equation 4.1 indicates the mathematical form of this relation. Projected migration is defined by a set of associated probabilities that demographers may assume to be constant, increasing, decreasing, or a function of variables which may be calculated via regression or other procedures. Migration, once derived, can be added to births, deaths, and the previous year's population in order to determine total population size.

Econometric Models

Econometric projections estimate population growth by considering economic factors. Assumptions about jobs, industrial growth, unemployment, the profile of the employment sector, retail sales, and other economic variables are incorporated in this type of model. The advantage of such an approach is that migration can be related to an area's economic attractors, such as jobs, high wages, and new construction. The disadvantage of this procedure is that it is more complex than demographic methods and does not necessarily provide greater accuracy (11).

Two Texas state agencies prepare population projections, the Texas Department of Health (TDH) and the Texas Department of Water Resources (TDWR). In addition, local planning entities and governments occasionally prepare their own projections or modify existing forecasts, presumably better to represent their particular needs or situation. This

---

The basic population equation:

$$P(t) = P(t - 1) + B - D + I - E \qquad (4.1)$$

where $P(t)$ = population in year t

    $P(t - 1)$ = population in year t - 1

      B = number of births in year t

      D = number of deaths in year t

      I = number of immigrants in year t

      E = number of emigrants in year t

---

section presents the methodologies and results of TDH and TDWR projections.

The Texas Department of Health and the Texas Department of Water Resources employ a modified cohort-component technique. In this approach each of the components--births, deaths, and migration--are projected separately for each five-year, ethnic/race, sex cohort. Both TDH and TDWR use sixteen age groups, three ethnic groups, and both sexes for each county, or a total of ninety-six cohorts.

TDWR Projections

The TDWR constructed 1980 population cohorts using 1970 U.S. census data and the number of births and deaths, as recorded by the Texas Department of Health (12). To estimate county births, the TDWR developed state-specific, age/ethnic cohort fertility rates based upon demographic data; these were adjusted toward the statewide mean "in order to conform with the standard statistical assumption of tendency towards the mean" (13). The TDWR death component was drawn from life tables covering the period of 1969-71, adjusted to reflect historical differences between U.S. and Texas death rates. These rates were then trended downward to account for the "historical trend of decreasing death rates" (14).

The TDWR then used least-squares econometric estimation to assess how various county characteristics affected migration from 1970 to 1980. This least-squares equation was then used to project county migration. Migration equations were developed for counties with populations (a) greater than or equal to 100,000, (b) between 5,000 and 99,999, and (c) less than 5,000.

Depending on the size-category of the county, independent variables to explain migration can include previous migration, the populations of neighboring counties, per capita income, housing starts, income change, 1970 population, the presence of a junior college, the January temperature, an instrumental (dummy) variable for counties bordering Mexico, travel expenditures, lignite deposits, or an instrumental variable for "hill country" counties. The resultant migration rates are then adjusted to conform to one of two independent estimates. One scenario assumes that Texas's high rate of migration during 1970-80 will continue through 1980-90. The second case assumes that migration will follow patterns of the period from 1950 to 1980, thus damping the unusually high rates of the 1970s. Projections beyond 1990 employ rates calculated for the 1980-90 period, but reduce the migration rate in a linear fashion to zero by the year 2100. The rationale for reducing this rate is that "it is not realistic to presume that the number of migrants into Texas will grow continuously at a constant rate through distant future time as a result of the relative attractiveness of Texas as a place to live and work" (15). The choice of a convergence year was "guided by independent projections of future manufacturing, mining, and agricultural activity in Texas prepared by the Texas Department of Water Resources. That is, the projected number of Texans in future decades is consistent with the number of available jobs and a reasonable proportion of the population participating in the work force" (16). The derived net migration is then added to projected births, deaths, and survivors to produce the TDWR forcasts.

The TDWR's highly systematic and economically based projection methodology reflects a meticulous and laudable concern for detail. The assumptions surrounding the construction of the component data (births and deaths) appear reasonable. However, the treatment of migration is guided by what might be termed "supply side" assumptions. It is all but explicitly stated that migration will be in direct proportion to the potential future jobs available in Texas. Certainly economic growth and development in manufacturing, mining, and agriculture will prove to be economic attractors for a labor force. But the other side of the equilibration, the "demand side," is that industrial relocation can depend on the supply of an adequately trained and available labor force. Whether the independently projected labor force is a target or an upper limit on population projections is not apparent. If a projected labor force is assumed to be the result of growth, an upward bias may be present.

TDH Projections

TDH similarly computes cohort-base projections from the base period 1970-80. Rather than use econometric techniques, the TDH "survives" the 1970 population to 1980 and calculates the difference between the survived population and the actual population. This computed growth rate is then applied to the population surviving from 1980 to 1990 to obtain estimated 1980-90 net migration. A sum of net migration, the survived population in 1990, and probable future births and deaths constitute the projected population for 1990. Projections for the year 2000 are obtained in the same manner, with 1980-90 as a base period and using the same survival and birth rates.

Survival rates are derived from abridged 1979 life tables of the white and nonwhite population. TDH uses white survival rates for the Hispanic population. As changes in estimated net migration are determined by the U.S. Bureau of the Census, TDH incorporates and adjusts the projections in its "Population Data System" accordingly. This system is a more straightforward demographic approach than TDWR's system of extensive multiple regression. The TDH methodology is based on two questionable assumptions, that survival ratios are constant over time and that migration, calculated from survived populations, is essentially static (17). One limitation of the first assumption--stable survival ratios--is that these fractions have not remained historically stable. The crude birth rate declined by 12 percent from 1970 to 1980, the crude death rate decreased by 6 percent, and the infant mortality rate dropped by 38 percent over the same decade (18).

Comparison of Projections

Tables 4.11 and 4.12 list the TDWR and TDH population projections from 1980 to the year 2000. Both agencies project greater future rates of growth than the fifty-year historical rate of 2.20 percent.

Is there much of a relation between the TDH and TDWR projections? If both use similar factors one might argue that the numbers should be linearly correlated; in other words, each could be a simple multiple of the other. The linear correlation coefficient of 0.187 does not support the assertion of a strong relationship. Not only are the population magnitudes dissimilar, but on occasion even the direction of change is different. For example, the TDH projects a virtually exploding 7.59 percent annual growth rate for Culberson County, while TDWR forecasts a modest decline of -0.07 percent.

The TDH consistently projects more rapid growth than TDWR in all border counties. While TDWR projects 3.05 percent annual growth in the region, TDH expects a 4.90 percent rate. TDH's statewide population growth rate is 68 percent larger than TDWR's, a significant difference. The change in population of the region, according to TDH, will be 166 percent--44 percent greater than TDWR estimates. This rate of growth will add two million inhabitants to Rio Grande/Río Bravo counties, or one million more than projected by TDWR. Moreover, TDH expects the Texas population in the year 2000 to be nearly 28 million persons, or 6.6 million more than TDWR.

The fact that these projections are so different raises questions about the validity of either forecast. The unprecedented levels of growth implied by the TDH projections raise doubts about the technical consistency and credibility of their model. Yet it is unclear whether the TDWR projections are better indicators of future population levels in the border region. Although everyone projects growth, the disparities among regional population projections suggest that the level of growth is a matter neither of agreement nor of certainty.

BORDER REGION POPULATION PROJECTIONS FOR MEXICO

Table 4.13 illustrates a comparison of estimates prepared by the Mexican *Secretaría de Programación y Presupuesto* (SPP) and the U.S. Census Bureau projections for Mexico and the United States. If Mexico's population grows at the U.S. Census Bureau's rate of 3.07 percent per year, it would double every twenty-three years. Increasing by 1.83 percent annually (the SPP's rate), the population would double in 38 years. At the U.S. growth rate of 0.60 percent, the number of residents will double only after 115 years.

The actual rate of growth of the population in Mexico from 1960 to 1980, according to the Mexican census, has been 3.12 percent annually. The growth rate of the four-state region over the same period has roughly tracked Mexico's level, increasing at a 3.10 percent annual rate. The fastest-growing border state, Nuevo León, has grown by an extraordinary 4.13 percent, doubling in size in only seventeen years.

Table 4.14 displays the magnitude and rate of growth in the border region and Mexico. It also lists extrapolations of the population in Mexico and the border region until the year 2000, assuming a continued high growth rate. According to these rough projections, this region will contain a population in excess of 11 million inhabitants shortly after 1990; over 15 million will reside there by the year 2000.

Table 4.15 lists the cities of Mexico and their projected populations, given two different growth rates, those of the the U.S. Bureau of the Census (3.07 percent) and the Mexican SPP (1.83 percent). The Census Bureau estimates that the border cities' population will exceed 2.6 million inhabitants; the SPP projects that about 2 million persons will inhabit Mexico's border cities.

There is no dispute that, given current growth rates, the Mexican border along the Rio Grande/Río Bravo will undergo tremendous demographic change within the next twenty years.

UNDOCUMENTED MIGRATION IN THE BORDERLANDS

A significant yet uncertain factor in population change in the Rio Grande/Río Bravo basin is the level of undocumented migration from Mexico to the United States. Although one source estimates that 21.2 percent of all Mexican legal and illegal immigrants in the U.S. reside in

Texas, neither the TDH nor the TDWR treat undocumented immigration as a variable in their projections (19). This oversight probably reflects the deficiency of information on immigration; as "undocumenteds" do not report their entry or presence in the U.S., reliable records simply do not exist. Sophisticated "guesstimating" techniques are the only available means of approximating the number of undocumented Mexican migrants.

The conventional wisdom places the stock of undocumented Mexican migrants in the United States between 400 thousand and 5.2 million (20). U.S. Bureau of the Census demographers reviewing studies for the Select Commission on Immigration and Refugee Policy concluded that the probable size of the "illegal resident population is almost certainly less than 3.0 million,. . . possibly only 1.5 to 2.5 million" (21). A more recent study based on differences in sex ratios between the United States and Mexico estimated that approximately 1.5 to 4 million undocumented migrants of Mexican origin resided in the United States in 1980 (22). Most demographers agree that such estimates are "instructive rather than definitive in nature" (23,24).

In summary, the clandestine nature of undocumented migration thwarts accurate representation of the number of undocumented migrants of Mexican origin in the Rio Grande/Río Bravo basin. One important lesson to be derived from this brief section is that, in the words of Siegel, Passell, and Robinson, "Researchers and policymakers will have to live with the fact that the number of illegal residents in the United States cannot be closely quantified. Therefore, policy options dependent on the size of this group must be evaluated in terms which recognize this uncertainty" (25).

CONCLUSIONS

The Texas/Mexico border region is expected to grow from its 1980 population of nearly 2.7 million persons to between 4.2 to 5.8 million inhabitants by the year 2000, depending on whose estimates are accepted. Such growth will affect the physical environment and infrastructure of the border region in profound ways.

The purpose of this study is to ascertain potential water quality and quantity issues for this region. This chapter of the report has demonstrated that the population of this region may more than double in the next two decades. As the number of inhabitants grows, so will the water quantity and quality demands. Unless efforts are made to provide water for the residents of the Rio Grande/Río Bravo basin, the Texas/Mexico border region may be physically, economically, and politically unable to cope with the thirsty demands of a rapidly increasing population.

As the population grows, communities will experience fiscal and economic stress as ever larger shares of the tax dollar are earmarked for water development, water treatment, and waste-water treatment. The more resources that are devoted to water projects, the less funds will be available for other government services such as roads, police, fire protection, and health care. In addition, a smaller pool of savings for investment purposes will dampen the prospects for private development, small business, and home mortgages. At some future time, either the physical limits of surface and subterranean water availability, or the revenue constraints of a community's taxing capacity will be reached. The well will have run dry. An integrated and balanced approach to water resource management planning is necessary to prevent the depletion of the region's water and financial resources.

REFERENCES

1. City of El Paso Department of Planning, Research, and Development, *A Demographic Analysis of El Paso City and County* (El Paso: City of El Paso, December 1980), pp. 6-7.

2. U.S. Bureau of the Census, *1980 Census of Population*, vol. 1, Characteristics of the Population, chap. A, part 45, (Washington, D.C.: U.S. Government Printing Office, March 1982).

3. Edmundo Victoria Mascorro, "Características del Desarrollo Económico de la Franja Fronteriza Norte de México," *Natural Resource Journal* 22, no.4 (October 1982): 827.

4. Michael V. Miller, *Economic Growth and Change along the U.S.-Mexico Border* (Austin: Bureau of Business Research, University of Texas at Austin, 1982), p. 3.

5. Ibid., p. 8.

6. All growth rates included in this chapter are continuous annual growth rates calculated by using the formula: $r = \ln(P(t)/P(o))/t$.

7. U.S. Bureau of the Census, *Census of Population and Housing, 1980*, Summary Tape File 3A (Washington, D.C.:   U.S. Government Printing Office, 1981).

8. U.S. Bureau of the Census, *USA Statistics in Brief 1981* (Washington, D.C.: U.S. Government Printing Office, 1982).

9. Ibid.

10. Ibid.

11. Judith Kunofsky, *Handbook on Population Projections* (San Francisco: Sierra Club, 1982), p. 76.

12. Texas Department of Water Resources, *Methods for Projecting Population for Texas Counties: 1990, 2000, 2010, 2020, and 2030* (Austin: TDWR, 1982), p. 4.

13. Ibid., p. 4.

14. Ibid., p. 5.

15. Ibid., p. 16.

16. Ibid.

17. Interview with Dr. Harrold Patterson, Texas Department of Health, February 6, 1983.

18. U.S. Bureau of the Census, *USA Statistics.*

19. Jacob S. Siegel, Jeffrey S. Passel, and J. Gregory Robinson of the U.S. Bureau of the Census for the Select Commission on Immigration and Refugee Policy, "Preliminary Review of the Number of Illegal Residents in the United States," in *U.S. Immigration Policy and the National Interest*, Appendix E to the Staff Report of the Select Commission on Immigration and Refugee Policy, Papers on Illegal Migration to the United States (Washington, D.C.: U.S. Government Printing Office, March 1981), pp. 15-29.

20. Ibid., p. 39.

21. Ibid., p. 33.

22. Frank D. Bean, Allan G. King, and Jeffrey S. Passel, *The Number of Undocumented Migrants of Mexican Origin in the United States: Sex Ratio-Based Estimates for 1980*, Paper 4.020, Texas Population Research Center Papers (Austin: Population Research Center, University of Texas at Austin, 1982), p. 11.

23. Ibid., p. 13.

24. Siegel, Passel, and Robinson, "Illegal Residents," p. 34.

25. Ibid.

CHAPTER 5

Water Use Along the
Rio Grande/Río Bravo

Even though the United States and Mexico have signed two treaties apportioning the waters of the Rio Grande/Río Bravo, the issue of *who uses how much water* along the Rio Grande/Río Bravo is not a straightforward matter. The 1906 treaty guarantees Mexico 60,000 acre-feet or 0.07 cubic kilometers (cu km) of water from the Elephant Butte Dam (1). The 1944 treaty allocates to Mexico an additional 1.5 million acre-feet (maf) or 1.85 cu km per year from the Colorado River and provides for dividing the river's waters below Fort Quitman (2). The allocation of those surface waters on the Texas side are a matter of legal record, as water use permits are issued by the Texas Department of Water Resources (TDWR).

Despite this legal framework the actual patterns of water use remain obscure. The uncertainty regarding "who uses how much water" stems from three factors: (a) ambiguity in determining what constitutes water use, (b) difficulties in measuring actual water utilization, and (c) lack of information regarding the future intentions of potential water consumers.

The purpose of this chapter is to make the best of this ambiguity by trying to quantify past and present utilization of water along the Rio Grande/Río Bravo. This section will define terms relating to water use.

The terminology of water resources planning can itself be a source of ambiguity. Reports contain various water-related terms such as *water use*, *withdrawal*, *water consumed*, or *water demand*. These terms are not interchangeable; none of them in fact exactly corresponds to the quantities reported by responsible water-resource agencies.

The terms *water use*, *water demand*, or *water requirement* are used to indicate a volume of water which is withdrawn for some use. The

85

amount of water brought through water intakes from either surface or ground sources is considered water withdrawn. If water is withdrawn once by an industry and then recycled a number of times in a manufacturing process, it may do the work of a greater amount of water. However, such reused water is counted only once as water withdrawn. Water returned to some surface stream, lake, or into an aquifer or ocean is considered a *return flow* that can be available for withdrawal by another user. Some water uses do not require withdrawal. Examples are water used for hydroelectric power generation, flow augmentation, or discharge into estuarine areas to maintain salinity gradients.

The term *consumption* or *water consumed* refers to that volume of water which is not available to be recycled as return flow as it has been transformed by evaporation, transpiration, percolation, or by incorporation into products, crops, livestock or people. Sewage is not a consumptive use, but rather a return flow.

Water withdrawn is the term that comes closest to corresponding to the water utilization data contained in this chapter. On the Texas side, TDWR estimates a portion of the water withdrawn from surface sources, groundwater, or return flows. Water utilization figures from the Mexican side are guesstimates or statements of intent; rarely do figures reflect actual water measurements. These inadequate data from both sides constitute the second source of uncertainty in evaluating water utilization along the Rio Grande/Rio Bravo.

In 1974 and 1977, water withdrawals for manufacturing were partly estimated by TDWR. TDWR gathered reported-use information by distributing voluntary surveys to manufacturing water users. The surveys requested total water withdrawals by manufacturer in the survey year. Municipal records of manufacturing water sales were also checked in an effort to substantiate the survey results (3). The levels of manufacturing water withdrawals published by TDWR in both 1974 and 1977 were determined by increasing these surveyed data by as much as 15 percent (4). TDWR added these estimates to account for the level of unreported withdrawals that it felt were missing from the surveyed information. Such use of administrative corrective factors was substantially reduced in 1980 (5).

Data regarding irrigation water withdrawals are subject to similar limitations. For example, in 1979 the Soil Conservation Service (SCS) collected information on irrigation water withdrawals. Accuracy of these reports varied among counties in relation to (a) the assigned individual analyst's knowledge of the area, (b) the amount and reliability of available water-use records, and (c) the total number of field observations made in a given area (6). SCS maintains that the 1979 irrigation water withdrawal figures are within 5 to 10 percent of the actual total (7).

Texas return-flow statistics, particularly TDWR's published irrigation return-flow levels, are also limited by the estimation process. Irrigation return flows are based solely on estimates, as no reports are mandated by TDWR in cases involving irrigation discharges (8).

Any analysis of Mexican water-withdrawal information is limited by the scarcity of published reports, especially in the area of groundwater withdrawals. This dearth of published information makes it difficult to draw firm conclusions on water-withdrawal problems within the study area. Some data, such as annual municipal water withdrawal rates, reflect estimates of actual withdrawals based on multiplying published daily withdrawal rates by 365. Extrapolation of yearly rates from daily municipal withdrawal rates creates some unique problems. For example, in a 1980 municipal study prepared for the *Secretaría Agricultura y Recursos Hidráulicos*, it was estimated that the cities of Nuevo Laredo, Piedras Negras, and Matamoros withdrew a combined total of 38.9 thousand acre-feet (48 million cubic meters) of industrial water from the Rio Grande/Río Bravo for domestic and industrial uses (9). In contrast, multiplying 1974 daily withdrawal rates by 365 and adding the 1974 withdrawals of these three cities yields a combined municipal use figure of approximately 46.3 thousand acre-feet (57.1 million cubic meters).

Discrepancies in the published data may be due to the techniques used for approximating estimated annual municipal withdrawal rates. For example, these 1974 data were collected on a one-day sample by a private Mexican consulting firm. The firm did not publish its methods of data collection or its sampling procedure. If samples were taken on a peak water-withdrawal day, extrapolation of these data would build bias into the figures. For instance, these procedures would generate 1974 municipal water withdrawals for the Mexican border cities that eclipse the 1980 figures. The reader should bear these data limitations in mind when examining this chapter.

PATTERNS OF WATER WITHDRAWAL

Irrigation water use has dominated water use in the Rio Grande/Río Bravo basin in the past. Fluctuations either upward or downward in agricultural water use have, as a consequence, affected major shifts in overall water-use patterns. Currently agricultural producers are contending with a rapidly expanding urban population within the region. As the basin's municipal population has increased, so has competition for available water. Other water-use sectors do not greatly influence water-use patterns, although significant increases in water withdrawals in any one of the other sectors (such as mining, livestock, or manufacturing) may in the future prove the difference between water supply adequacy or scarcity.

Total estimated water withdrawals within the Rio Grande/Río Bravo basin, circa 1974-76, are presented in Table 5.1. Figure 5.1 graphs water withdrawals along the Texas side of the border for 1974, 1977, and 1980. Note the consistent decline in groundwater utilization that occurred over this six-year period. Surface water withdrawals, on the other hand, fluctuated between 1974 and 1980. If it is assumed that estimated Mexican water use remained fairly stable from 1974 to 1980, then the range for total basin water withdrawals from 1974 to 1980 would be between 5.0 and 7.0 maf (between 6.17 and 8.63 cu km) per year. The actual total is probably closer to the latter figure, as published Mexican information regarding water withdrawals does not contain information on several water-use sectors and incompletely reports groundwater withdrawals. Given this unreliability and lack of hard information, these data neither confirm nor deny published assertions that there is little if any surface water available for uses outside the immediate Rio Grande/Río Bravo area (10).

Figure 5.2 illustrates annual water-use patterns in each of the thirty-three Texas counties and four Mexican states. These withdrawals are aggregated into upper, middle, and lower reaches and listed in Table 5.2. Twenty-three of the thirty-three counties used less than 50,000 acre-feet (0.06 cu km) of water annually. The major water-using counties were El Páso and Reeves in the upper reach, and Hidalgo and Cameron in the lower. In the upper reach, six of its fifteen counties exceeded water use levels in excess of 50,000 acre-feet (.06 cu km) In contrast, only Maverick, Uvalde, and Zavala surpassed this level in the middle reach. The lower reach contains the region's two most intensive water-using counties--Hidalgo and Cameron. The lower reach's remaining seven counties collectively used less than 50,000 acre-feet (.06 cu km) of water annually.

If only the Texas side of the border is considered, the disaggregated water-withdrawal data for 1974, 1977, and 1980 reveal some important trends (see Tables 5.3 and 5.4). Water withdrawals consistently declined from 1974 to 1980 in the middle reach. The TDWR data suggest that future declines may continue, although fluctuations in water withdrawals may be caused by a variety of factors.

The intense agricultural activity in the Rio Grande/Río Bravo basin accounts for irrigation's primary role in the water-use picture from 1974 to 1980. Table 5.4 lists the estimated volumes of water withdrawn by sector for each of the three reaches. When examining this table, it

(continued on page 91)

Figure 5.1

**Reported Water Withdrawals along the Texas Side of the
Rio Grande/Río Bravo Basin for 1974, 1977, and 1980
(1,000s of acre-feet)**

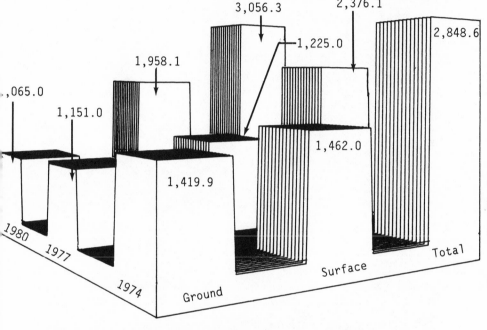

Sources: Texas Department of Water Resources, Planning and Development Division, Economics Section, "Fresh Water Use in Texas: 1974
and 1977," Austin, 1982. (Unpublished Computer Printout.)

Texas Department of Water Resources, Planning and Development Division, Economics Section, "1980 Water Use by Category and Source,"
Austin, 1982. (Unpublished Computer Printout.)

Figures are rounded to the nearest hundred acre-feet.

Figure 5.2

Annual Water Withdrawals in the Rio Grande/Río Bravo Basin

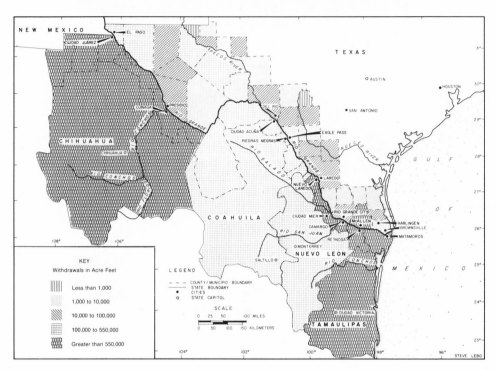

Sources: See References and notes for Table 5.1.

is important to keep in mind that Mexican water-use statistics include only irrigation and municipal water withdrawals.

A comparison of the agricultural withdrawals of Table 5.4 with the total withdrawals of Tables 5.2 and 5.3 indicates the importance of irrigation in the basin's water-utilization picture. In the past agricultural water withdrawals have accounted for approximately 88 percent of the Texas basin's total water withdrawals. If the upper and lower reach irrigation-use figures are combined, they constitute over 87 percent of all water withdrawals in the basin (see Table 5.4).

Return flows provide an additional, although indeterminant, source of water to augment ground and surface water. Municipal and industrial return-flow data from the Texas border region are listed in Table 5.5 by reach. Project members were unable to obtain statistical data regarding the volume of Mexican return flow. The TDWR maintains estimates of the amount of irrigation return flow that is reused. It does not report information as to the volume of municipal or industrial water return flows that are reused.

Water scarcity is a present reality all along the Rio Grande/Río Bravo. Each reach is either currently experiencing or projected to experience some water shortfall or scarcity problem in the not too distant future. In the El Paso/Ciudad Juárez area, the dearth of surface supplies has encouraged the cities to pump groundwater rapidly from a contiguous aquifer. As a result, water levels in the aquifer are being rapidly depleted. As the depletion accelerates, quality problems increase; consequently, the area is suffering from potential shortages of water and increased quality degradation in remaining supplies. A similar phenomenon is occurring in the Trans-Pecos irrigation region in West Texas. Irrigation costs are rising so fast that some agricultural producers may seek alternative land uses. Increased pumping costs are also playing havoc with irrigation in the counties of Uvalde and Zavala. Groundwater dependence in this area is a root of this problem in both irrigation areas. Water-availability problems plague municipalities in the lower and middle reaches. The TDWR projects that the cities that depend on water supplies from Falcon and Amistad reservoirs may experience shortages by 1985 (11). The lower reach in both Texas and Mexico presently overuses available surface water supplies. Each of these problem areas will be covered in greater detail later in this chapter. The next section disaggregates water use by sector, source, and reach.

ESTIMATED WATER USE

This section characterizes patterns of water use, such as the types of users and the volumes they utilize. Water consumers are categorized in terms of the nation of origin (U.S. or Mexico), the portion of the river basin (upper, middle, or lower reach), the county or municipio unit (33 Texas counties and 4 Mexican states), the water use (agricultural, municipal, industrial, or manufacturing sectors), and the source of water supply (surface water, groundwater, or return flows).

Although groundwater has become less important over time along the Texas side of the border (see Figure 5.1), the level of groundwater dependency varies by reach. As indicated in Table 5.6, groundwater availability can mean the difference between adequate supplies of water and critical shortages in some cases. Although overall groundwater use as a percentage of total use declined in the Texas border region, the upper and middle reaches depend upon groundwater.

Mexican groundwater dependency is difficult to determine given the dearth of published information. According to one analyst, Mexico's agricultural and municipal sectors have in the past strongly relied on groundwater to meet water demands (12). Unfortunately, evidence to substantiate such an assertion is not available from published sources.

Municipal

Municipal water use has risen steadily within the border area (see Figure 5.3). Along the Texas side of the border, municipal water withdrawals increased over 31 percent between 1974 and 1980 (see Figure 5.3). Although municipal water withdrawals along the Mexican side of the border region are likely to have increased, there is a scarcity of comparative water use data. Table 5.7 lists municipal water use in 1974 for the most populous Mexican border cities.

Figure 5.4 illustrates the six major contiguous cities in the Río Grande/Río Bravo basin. The Texas cities in the middle and lower reaches used little if any groundwater. The corresponding Mexican municipalities had relatively greater groundwater dependence.

The Texas cities identified in Figure 5.4 (El Paso, McAllen, Laredo, Eagle Pass, Brownsville, and Presidio) have historically dominated

(continued on page 95)

Figure 5.3

**Reported Municipal Water Withdrawals along the
Texas Side of the Rio Grande/Río Bravo Basin for
1974, 1977, and 1980**

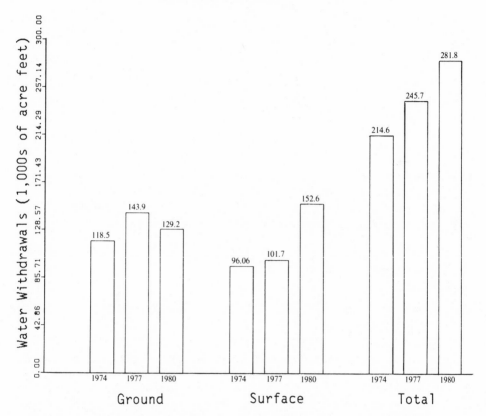

Sources: Texas Department of Water Resources, Planning and Development Division, Economics Section, "Fresh Water Use in Texas: 1974, 1977, and 1980," Austin, 1982. (Unpublished Computer Printout.)

Texas Department of Water Resources, Planning and Development Division, Economics Section, "1980 Water Use by Category and Source," Austin, 1982. (Unpublished Computer Printout.)

Figure 5.4

The Major Water-Withdrawing Municipalities
in the Rio Grande/Río Bravo Basin

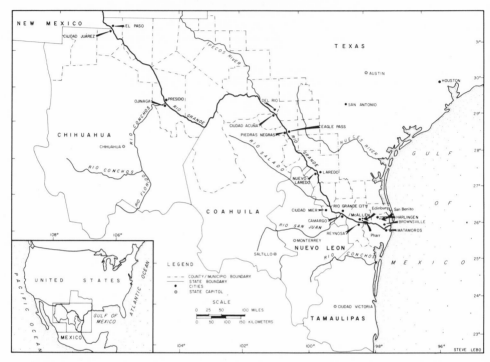

Sources: Texas Department of Water Resources, Planning and Development Division, Economics Section, "Fresh Water Use in Texas: 1974,
1977, and 1980," Austin, 1982. (Unpublished Computer Printout.)

Texas Department of Water Resources, Planning and Development Division, Economics Section, "1980 Water Use by Category and Source,"
Austin, 1982. (Unpublished Computer Printout.)

municipal water use in the Texas border region. In 1974, these six collectively withdrew 60 percent of the total water used for municipal purposes in the region. El Paso's share of municipal water use declined from 1970 to 1980. McAllen's share remained relatively stable. The other four cities experienced increases in their share of total municipal water use (see Table 5.8).

Table 5.9 identifies the four major standard metropolitan statistical areas (SMSAs) in the region. Together they accounted for 65 percent of the Texas 1980 municipal water withdrawals and 66 percent of the municipal population in the Rio Grande/Río Bravo basin.

Each of the six major Texas border cities experienced an absolute increase in municipal water use over the 1970-80 decade. This pattern of municipal water withdrawal growth probably occurred in Mexico as well.' As a general rule, per capita water use in Mexican cities tends to be lower than rates found in comparable American municipalities (see Table 5.8). Mexican cities tend to have both lower rates of domestic water use and fewer water-intensive industries (13). Table 5.10 presents the figures for 1974 municipal water use in the six contiguous border cities.

As noted above, municipal water-use rates grew in the Texas municipal areas by about 24 percent between 1974 to 1980. Although historical data regarding Mexican municipal water withdrawals are sparse, it is possible to assume that a city's total water use would increase in proportion to its population growth. Table 5.11 shows the rapidity with which some border-city populations have grown within the last two decades.

The inadequacy of most of the existing water service systems in Mexican border cities may be a major factor in holding down use. Indeed, Table 5.7 shows that the percentage of the population served with potable water varied from a high of 92 percent in Ciudad Juárez to a low of 66 percent in Reynosa. President Lopez-Portillo's National Urban Development Plan placed a priority on expanding potable water supplies in the border region (14). If successful, such increased supplies would provide Mexican border cities with access to additional water supplies (15).

Industrial growth may further contribute to increased municipal water withdrawals in the Mexican border municipalities. In the early 1960s, the Border Industries Program (BIP) was initiated in Mexico to "foster the creation of jobs through the installation of in-bond plants which will be involved in assembling and processing American products" (16). Since 1965, the BIP has received heavy promotion from the Mexican government (17). By 1977, the industries from all of Mexico's border states accounted for 6.3 percent of its balance-of-payments income (18).

The success of the BIP may be reflected in the per capita income of the Mexican states bordering Texas. As of 1965, three of the four border states were considered wealthy relative to Mexico's interior states; the fourth was considered to be a relatively middle-income state (see Table 5.12). At least one analyst contends that these four states are increasing in affluence relative to the interior states (19). Increasing incomes may be associated with greater demand for municipal water as the population begins to acquire water-intensive appliances (20).

If these areas continue to grow in size, serious water difficulties between the contiguous border cities may be unavoidable. Critical shortages that have been experienced previously in some municipalities may be exacerbated. For example, in the summer of 1982 four children died of dehydration and hundreds were hospitalized when Juárez water trucks failed to meet the water needs of its poorer neighborhoods (21).

Water availability is just one of the water-use problems in the El Paso/Ciudad Juárez sector. There is also a rapid depletion of the Hueco Bolson aquifer that serves both these cities. Heavy pumping is causing saline water encroachments, which could degrade the area's principal water source (22). As a result, water difficulties could occur. The city of El Paso is currently worried that existing groundwater supplies may be deficient by 1995 (23). A suit was recently won by El Paso against New Mexico to allow unrestricted interbasin transfers of New Mexico's groundwater to quench El Paso's thirst.

Future water problems are also anticipated in those municipalities served by Lake Amistad and Lake Falcon, or all of the municipalities located within the counties running from Val Verde along the Rio Grande/Río Bravo to Cameron. (This area includes the sister cities of Eagle Pass/Piedras Negras, Laredo/Nuevo Laredo, McAllen/Reynosa, and Brownsville/Matamoros). Given present water utilization rates of each of the major SMSAs in this area, the TDWR estimates that the Texas municipalities may experience surface water shortages beginning in 1985 or 1990 (24). One might suspect similar water shortage problems in the twin cities along the Mexican side of the basin.

Industrial

Industrial water use for mining (fuels and nonfuels), livestock, and manufacturing rose significantly in the Texas study area between 1974 and 1980. Table 5.13 shows the trend by reach during the 1974-80 period. At first glance these data suggest a continuing decline in industrial water use in both the middle and lower reaches, with a possible stabilization in the upper reach. A projected tar-sands development project in the middle reach could significantly increase industrial water withdrawals if it should become an operational industry. The next subsections consider water withdrawals in specific industrial sectors of the Texas study area.

Oil and Gas

A number of natural gas and oil fields are concentrated in both the lower and upper reaches of the Rio Grande/Río Bravo basin. Figure 5.5 illustrates the major mineral fuels located in the state of Texas. Deposits of subsurface or near-surface lignite are deposited in the middle reach. Table 5.14 lists water withdrawals for fuels mining in the Texas border region during 1974 and 1977. Present oil and gas industry water withdrawals are relatively inconsequential. The projected tar-sands development project in the middle reach could substantially increase water withdrawals in this sector.

Mining (Metallic, Nonmetallic, and Fuels)

Figure 5.6 illustrates the types of minerals found within the Texas Río Grande/Río Bravo basin. Table 5.15 disaggregates mining water withdrawals by reach and subsector for the Texas study region. These data seem to indicate that (a) nonmetallic mining is a major water-use sector and (b) the volume of water withdrawn in the metallic mining sector is not great. Water withdrawals for nonmetallic mining activities jumped dramatically between 1974 and 1977. Impressive increases occurred in the nonmetallic mining sector. These activities, which are located principally in Ward, Culberson, Pecos, and Reeves counties (see Figure 5.7), generated a mining water withdrawal increase of more than 588 percent between 1974 and 1977.

Figure 5.5

Mineral Fuels Located along the Texas Side
of the Rio Grande/Río Bravo Basin

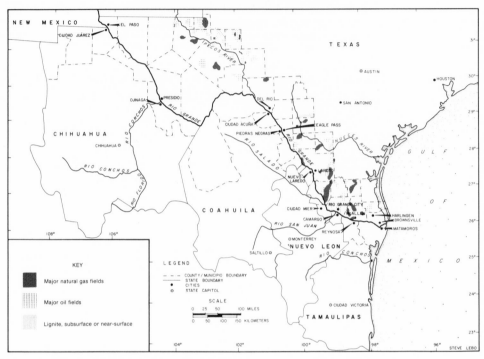

Sources: Texas Water Development Board, "Continuing Water Resources Planning and Development for Texas: Phase I," Volume I, Austin, May 1977, p. II-54. (Draft.)

Figure 5.6

Minerals Located along the Texas Side
of the Rio Grande/Río Bravo Basin

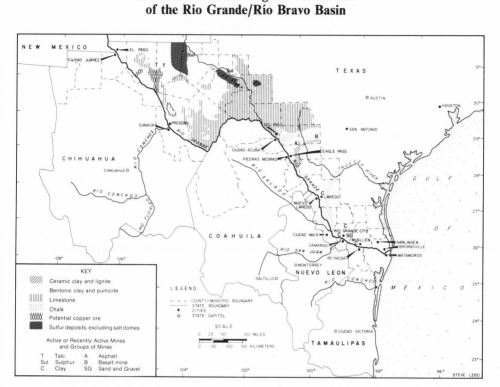

Source: Texas Water Development Board, "Continuing Water Resources Planning and Development for Texas: Phase I," Volume I, Austin, May 1977, p. II-52. (Draft.)

Figure 5.7

Mining Water Withdrawals along the Texas Side
of the Rio Grande/Río Bravo Basin, 1977
(acre-feet)

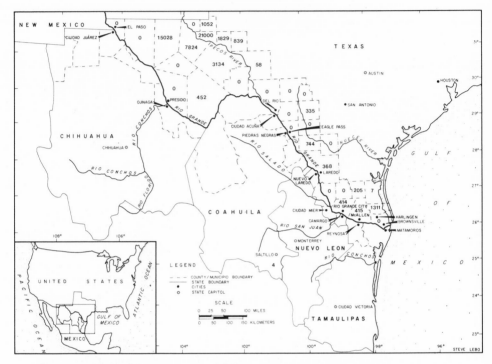

Source: Texas Department of Water Resources, Planning and Development Division, Economics Section, "Fresh Water Use in Texas: 1974 and 1977," Austin, 1982. (Unpublished Computer Printout.)

Along the Mexican side of the basin, there are indications that coal mining may increase in the state of Tamaulipas. If so, one can expect an increased demand for water in that sector.

Livestock

Every Texas county within the study region reported some water withdrawals for livestock. In the past livestock water withdrawals have been heaviest in Webb County. Livestock water use ranged from a high of 2,300 acre-feet in Webb (0.28 billion cubic meters) to a low of 100 acre-feet in Loving (0.12 million cubic meters). Figure 5.8 and Table 5.16 illustrate that water withdrawals for the livestock industry have remained relatively stable in the upper Texas reach, but not in the middle and lower reaches. It is not known why withdrawals have declined in the middle and lower reaches between 1974 and 1980.

Manufacturing

Manufacturing water activities made up the fourth largest water-using sector within the Texas study region. El Paso consistently ranked as the primary manufacturing county in terms of water withdrawals, with Zavala, Cameron, and Hidalgo following closely behind. In 1980, these four counties accounted for 96 percent of the total manufacturing water withdrawn in the region. Oil refining, located primarily in El Paso County, is the principal water-using manufacturing activity in the region. During the period of 1974 to 1980, oil refining used about 25 percent of the water withdrawn by all manufacturing categories along the entire U.S. stretch of the Rio Grande/Río Bravo border (25).

Figure 5.9 identifies the reported historical trends associated with manufacturing water use. It is not clear whether the 29 percent decline from 1977 to 1980 may in fact be due to TDWR estimation techniques or whether it represents an actual decline. Manufacturing water withdrawal data are based upon voluntary surveys. TDWR modified the survey results by adding estimated withdrawals to the reported information. As much as 15 percent of TDWR's final published 1974 and

(continued on page 104)

Figure 5.8

**Reported Livestock Water Withdrawals along the
Texas Side of the Rio Grande/Río Bravo Basin for
1974, 1977, and 1980**

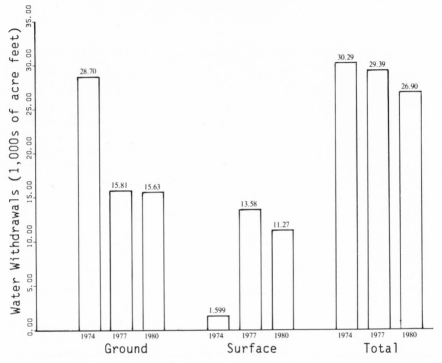

Source: Texas Department of Water ,Resources, Planning and Development Division, Economics Section, "Fresh Water Use in Texas: 1974,
1977, and 1980," Austin, 1982. (Unpublished Computer Printout.)

Texas Department of Water Resources, Planning and Development Division, Economics Section, "1980 Water Use by Category and Source,"
Austin, 1982. (Unpublished Computer Printout.)

Figure 5.9

Reported Manufacturing Water Withdrawals along the Texas Side of the Rio Grande/Río Bravo Basin for 1974, 1977, and 1980

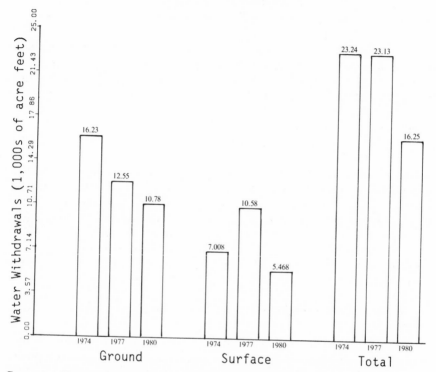

Source: Texas Department of Water Resources, Planning and Development Division, Economics Section, "Fresh Water Use in Texas: 1974, 1977, and 1980," Austin, 1982. (Unpublished Computer Printout.)

Texas Department of Water Resources, Planning and Development Division, Economics Section, "1980 Water Use by Category and Source," Austin, 1982. (Unpublished Computer Printout.)

1977 totals for manufacturing water withdrawals was "estimated" by TDWR. These withdrawal estimates are based on specific coefficients of water use for each standard industrial classification code (SIC). In 1980 the TDWR did not add as much estimated information to the survey results (26).

Table 5.17 summarizes these data on total manufacturing water use by reach. Note that each reach experienced major apparent declines from 1974 to 1980. Some rather dramatic losses in reported manufacturing water use occurred over the 1974-80 period in some counties. For example, eleven of the fifteen counties with no reported manufacturing water withdrawals in 1980 had withdrawals in 1974. The most dramatic decrease occurred in Real County. Real's 1974 manufacturing water withdrawal level of 565.7 thousand acre-feet (0.70 cu km) completely vanished in TDWR's 1980 statistics.

Table 5.18 reports the manufacturing sectors within Texas by two-digit Standard Industrial Classification (SIC) codes and the percent of water withdrawn in each industry for types of industrial water needs: process water, cooling, boiler feed, or other. Five of these industries-- food and kindred products, paper and allied products, chemical and allied products, petroleum refining, and primary metals--accounted for approximately 85 percent of the Texas total industrial water use in 1971 (27). In 1974 these five industrial categories withdrew approximately 80 percent of the total manufacturing water used within the study region (28). In 1980, a year when TDWR did little estimation of water withdrawals, these same five manufacturing types accounted for a total of approximately 93 percent of the reported manufacturing water used (29).

Project members were unable to obtain sufficient manufacturing water-use data from Mexico to determine overall water consumption patterns. However, the information available provides a general idea of the type and location of the major Mexican border industries. Five principal industrial plants operate within the basin of the Rio Grande/Río Bravo (30). These are La Domínica, S.A.; Altos Hornos de Mexico, S.A.; Refinería de Petróleos Mexicanos, S.A.; Celanese Mexicanos, S.A.; and Química Flour, S.A. Three of the five major border manufacturing plants are in the state of Tamaulipas and two are in the state of Coahuila. When considering water-withdrawal information on these plants, note that each of them relies exclusively on groundwater supplies.

La Domínica, S.A., is located near La Linda, in the state of Coahuila. It lies approximately 205 miles (330 km) downstream from the convergence of the Conchos and Rio Grande rivers at Ojinaga, in the state of Chihuahua, and 199 miles (320 kilometers) above Ciudad Acuña. Fluorite ore deposits are present within this area. These ore deposits contain a variety of impurities which La Domínica, S.A., filters out to obtain calcium fluorite. Its principal water source is a well located on the right bank of the Rio Grande/Río Bravo (31).

Altos Hornos de Mexico, S.A., is located in Piedras Negras. It initially opened in 1938 to produce cast iron and steel. Altos Hornos produces 80,600 tons of cast iron and steel annually, requiring iron, natural gas (121 cubic meters/ton), oil (52 liters/ton), and water (28 liters/ton) (32). The water used in production comes from deep wells located near the plant (33).

The Refinería de Petróleos Mexicanos, S.A., is located in Reynosa, in the state of Tamaulipas. It is one of Mexico's oldest and most important refineries. It first began operation in 1951 and includes a series of plants. The feed stocks are a low-sulphur crude oil and natural gas, which are extracted in the region. Water is derived from wells located on company-owned land. A polyethylene plant, which is part of the production process, is located approximately 6.2 miles (9.92 kilometers) west of Reynosa, outside of the immediate plant area (34).

Celanese Mexicanos, S.A., began operations in 1958, in the vicinity of Río Bravo City, in the state of Tamaulipas. Its principal products are explosives, paints, soaps, paper, and additional products that use synthetic fibers as principal raw materials. Floss cotton is the primary raw material used in production. Water for plant operations comes from three deep wells (35).

Química Flour, S.A., is the final major industrial water user within the Rio Grande/Río Bravo region of Mexico. Acid is its principal product. Raw materials come from residuals of the production process of La Domínica, S.A. Well water constitutes the principal water source (36).

A number of factors, including (a) the reliability of TDWR's withdrawal estimates, (b) the dearth of published Mexican data, (c) uncertainty regarding the Mexican economy, and (d) possible developments along the border make it difficult to draw any firm inferences regarding trends in future manufacturing water withdrawals.

Irrigation

Water withdrawals for irrigation represent the single largest use of water within the Rio Grande/Río Bravo basin. Figure 5.10 shows the amount of water used for irrigation by source. Groundwater withdrawals declined between 1974 and 1980; surface supplies increased.

Figure 5.10

**Reported Irrigation Water Withdrawals along the
Texas Side of the Rio Grande/Río Bravo Basin for
1974, 1977, and 1980**

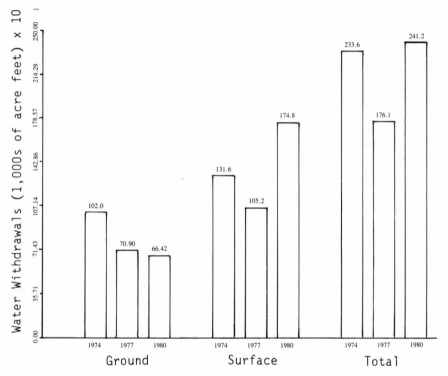

Source: Texas Department of Water Resources, Planning and Development Division, Economics Section, "Fresh Water Use in Texas: 1974 and 1977," Austin, 1982. (Unpublished Computer Printout.)

Texas Department of Water Resources, Planning and Development Division. Economics Section, "1980 Water Use by Category and Source," Austin, 1982. (Unpublished Computer Printout.)

Table 5.19 illustrates this pattern of fluctuating irrigation water withdrawals.

Tables 5.19 and 5.20 disaggregate the pattern of historical agricultural water withdrawals along both the Texas and Mexican sides of the border. The decline between 1974 and 1980 in Texas irrigation water use in the upper and middle reaches is not easy to explain; it may be due to any of a number of possible factors.

Both sides of the border contain areas of intense agricultural production, as illustrated in Figure 5.11. Table 5.21 lists information regarding the Mexican irrigation areas. It should be noted that irrigation water use in each of the Mexican irrigation districts has increased substantially since 1966. In some areas the volume of irrigation withdrawals more than doubled during this ten year period. Only the Acuña Falcon irrigation district experienced a decline between 1966 and 1976.

Similar historical information for each of the four primary Texas border irrigation regions is contained in Figures 5.12 (a-d) to 5.14 (a-d). These four areas accounted for over 89 percent of the region's total irrigation water withdrawals in 1974. Irrigated acreage has remained relatively stable in the El Paso Valley and Lower Rio Grande Valley since 1958. Irrigated acreage in the Trans-Pecos Region peaked in 1964 and has declined steadily since. The Middle Rio Grande Valley experienced a similar pattern, although there the irrigated acreage peaked in 1969. It has yet to reach its 1958 acreage level, however. In each of the regions except for the Trans-Pecos, the irrigation decline is only relative, as it was preceded by major increases in irrigated acreage.

Irrigated acreage has declined, but total irrigation withdrawals of water in these areas have decreased at a faster rate. Indeed, since 1958, the four regions underwent a collective decline of about 24 percent in total irrigated acreage, while water withdrawals fell by more than 34 percent.

In each of the areas except the Lower Rio Grande Valley, the amounts of water withdrawn for irrigation varied with the total number of irrigated acres in each region. Part of the decline may be explained by differing rainfall patterns (37). Indeed, approximately 140,000 acres or 56,658 hectares (ha) in both the El Paso Valley and Trans-Pecos Region that were equipped for irrigation and had previously been irrigated were not irrigated in 1979 (38). Over 90 percent of this total was located in the Trans-Pecos Region (39).

The declines in irrigation may reflect groundwater dependence. Farmers in the Trans-Pecos Region rely upon groundwater; an average of 92 percent of its irrigation water came from groundwater sources (40). Irrigation water is drawn primarily from surface or a combination of surface-ground sources in each of the other three areas (see Figure 5.12).

Figure 5.11

Major Irrigation Areas Located Within the
Rio Grande/Río Bravo Basin

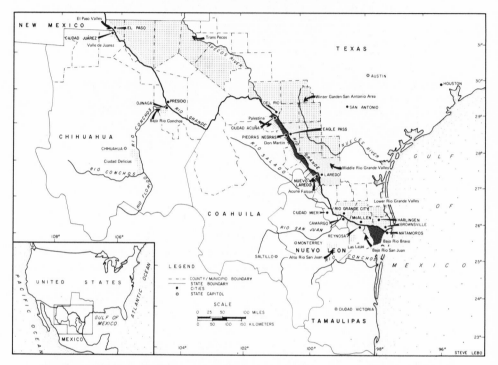

Sources: Felipe Ochoa y Asociados, S.C., *Estudio de la Calidad del Agua en la Cuenca del Río Bravo* (Mexico, D.F.: Secretaría de Agricultura y Recursos Hidráulicos, September 1978). (Limited Distribution Document.)

Texas Department of Water Resources, *Inventories of Irrigation in Texas: 1958, 1964, 1969, 1974, and 1979*, Report 263 (Austin: TDWR, October 1981).

Figure 5.12

Irrigation Water Withdrawals, 1958-1979

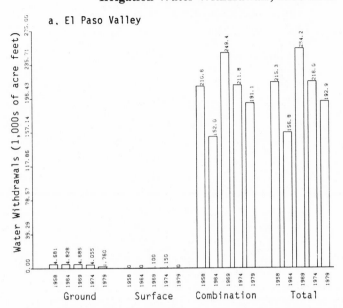

Source: Texas Department of Water Resources, *Inventories of Irrigation in Texas: 1958, 1964, 1969, 1974, and 1979,* Report 263 (Austin: TDWR, October 1981).

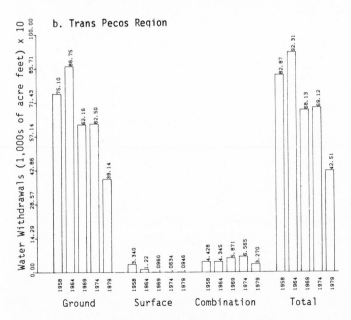

Source: Texas Department of Water Resources, *Inventories of Irrigation in Texas: 1958, 1964, 1969, 1974, and 1979,* Report 263 (Austin: TDWR, October 1981).

Figure 5.12 (Continued)

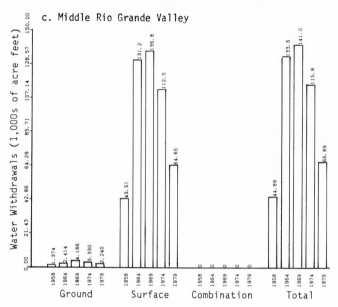

Source: Texas Department of Water Resources, *Inventories of Irrigation in Texas: 1958, 1964, 1969, 1974, and 1979,* Report 263 (Austin: TDWR, October 1981).

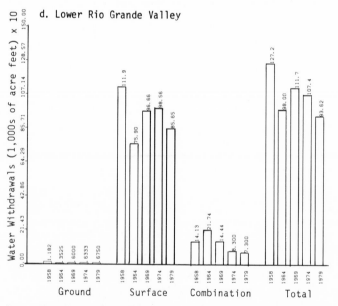

Source: Texas Department of Water Resources, *Inventories of Irrigation in Texas: 1958, 1964, 1969, 1974, and 1979,* Report 263 (Austin: TDWR, October 1981).

Figure 5.13

Acreages of Major Irrigated Crops, 1958-1979

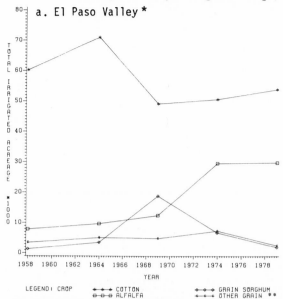

a. El Paso Valley *

Source: Texas Department of Water Resources, *Inventories of Irrigation in Texas: 1958, 1964, 1969, 1974, and 1979,* Report 263 (Austin: TDWR, October 1981). *Includes irrigated acreage for El Paso and Hudspeth counties. **Does not include grain sorghum, rice, wheat, or corn.

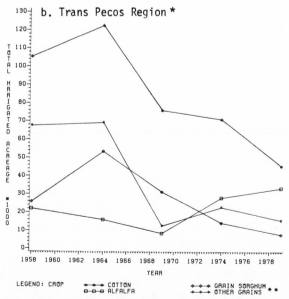

b. Trans Pecos Region *

Source: Texas Department of Water Resources, *Inventories of Irrigation in Texas: 1958, 1964, 1969, 1974, and 1979,* Report 263 (Austin: TDWR, October 1981). *Includes total irrigated acreage for Hudspeth, Culberson, Reeves, Ward, and Pecos counties. **Does not include grain sorghum, rice, wheat, or corn.

Figure 5.13 (Continued)

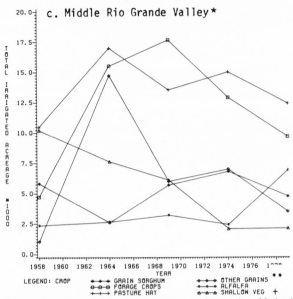

Source: Texas Department of Water Resources, *Inventories of Irrigation in Texas: 1958, 1964, 1969, 1974, and 1979,* Report 263 (Austin: TDWR, October 1981). *Includes total irrigated acreage for Maverick and Webb counties. **Does not include grain sorghum, rice, wheat, or corn. + For 1958 includes deep-root vegetables.

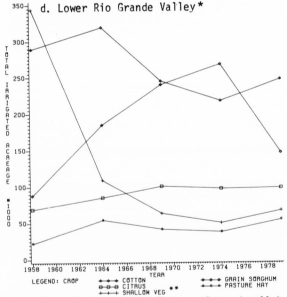

Source: Texas Department of Water Resources, *Inventories of Irrigation in Texas: 1958, 1964, 1969, 1974, and 1979,* Report 263 (Austin: TDWR, October 1981). *Includes total irrigated acreage for Willacy, Cameron, Starr, and Hidalgo counties. **For 1958 includes deep-root vegetables.

Figure 5.14

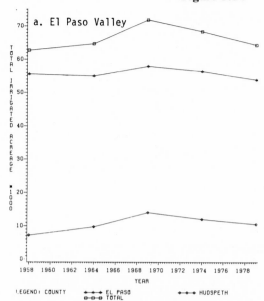

Source: Texas Department of Water Resources, *Inventories of Irrigation in Texas: 1958, 1964, 1969, 1974, and 1979,* Report 263 (Austin: TDWR, October 1981).

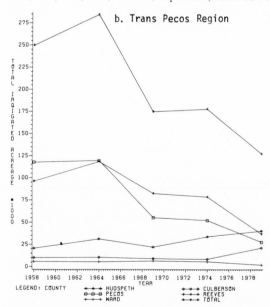

Source: Texas Department of Water Resources, *Inventories of Irrigation in Texas: 1958, 1964, 1969, 1974, and 1979,* Report 263 (Austin: TDWR, October 1981).

Figure 5.14 (Continued)

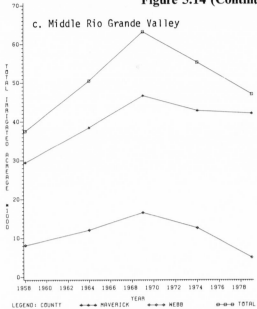

c. Middle Rio Grande Valley

Source: Texas Department of Water Resources, *Inventories of Irrigation in Texas: 1958, 1964, 1969, 1974, and 1979,* Report 263 (Austin: TDWR, October 1981).

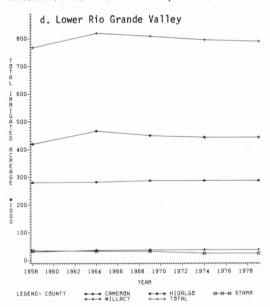

d. Lower Rio Grande Valley

Source: Texas Department of Water Resources, *Inventories of Irrigation in Texas: 1958, 1964, 1969, 1974, and 1979,* Report 263 (Austin: TDWR, October 1981).

Some analysts contend that groundwater is emerging as an important source for irrigation in Mexico as available surface sources in some areas are already overextended (41). For example, the 60,000 acre-feet (0.07 cu km) of water allotted to Mexico from the Rio Grande in the Juárez region are used downriver to grow cotton and pecans (42). Border area agricultural production tends to be large-scale, market-oriented, and intensive; the subsistence farming that characterizes much of Mexico's agricultural production is rare (43).

Increased agricultural expansion is a policy of the Mexican government. In 1979 the administration of President Lopez-Portillo initiated the *Sistema Alimentaria Mexicano* (SAM). The SAM had two goals: (a) to increase the rate of agricultural production in Mexico by 4 percent annually between 1980 and 1982 and (b) to attain self-sufficiency in corn and beans by 1982 and in basic staples by 1985. The SAM proposed that 4.7 million hectares (11.6 million acres) of new and reclaimed Mexican land (especially in the states of Chihuahua and Tamaulipas) be put into production by 1982. In these states, the Mexican government has pursued a policy of converting land once used for livestock production to agriculture (44).

Expansion of water use for agriculture may be difficult in some areas--especially in the Trans-Pecos, the Winter Garden-San Antonio area (around Dimmit and Zavala), and in the lower reach of the Rio Grande/Río Bravo basin. Dependence on groundwater limits agricultural expansion in the Trans-Pecos and Winter Garden-San Antonio irrigation regions. Pumpage exceeds natural recharge and as a result saline encroachment is occurring, causing the remaining aquifer water to become increasingly saline (45). The quality of the remaining groundwater is deteriorating as pumping costs increase. Figure 5.13 illustrates the approximate acreages of irrigated crops in each of these four areas. Alfalfa is on the verge of becoming the preeminent crop in terms of irrigated acreage in three of the four areas. In the Lower Rio Grande Valley, cotton continues to predominate. Cropping patterns are important because in large part they determine water needs. For example, farmers in the El Paso Valley are attempting to remain financially solvent by increasing their reliance on salt-tolerant crops, such as cotton (46).

In cases of serious groundwater decline, pumping costs can preclude irrigation sooner than either increased salinity or depletion (47). Groundwater extraction costs are determined by (a) the height the water must be raised (commonly known as the pump lift); (b) the quantity of water which is pumped; and (c) the quantity of energy used to obtain the necessary amount of water. As water levels decline and the required lift increases, pumping yields fewer gallons per minute. Hence, irrigators must pump longer to obtain the same amount of water. Energy costs for pumping are directly affected by each of these three factors.

Between 1973 and 1979 the energy costs for electric pumping of one acre-foot of groundwater in Texas have gone from $4.58 to $9.95. Natural gas pumping of one acre-foot of groundwater costs relatively less, but natural-gas pumping costs in Texas also increased during this period from $1.50 to $6.00 per acre-foot. These increased costs are the most burdensome in those areas having higher than normal pumping lifts (defined as exceeding 200 ft or 61 m) and rapid rates of aquifer decline (more than 3 ft or 0.9 m annually) (48).

The pumping lift for groundwater in the Trans-Pecos Region is now over 225 feet (69 m) and groundwater levels are declining at a rate of approximately 4 feet per year (1.2 m) (49). Pecos and Reeves counties have been the hardest hit by increased energy costs. Farming in these areas has become economically uncertain; as a consequence, many farmers have moved out of agriculture and into livestock production (50).

Groundwater irrigation in the region around the Winter Garden-San Antonio area is also plagued by pump lifts of about 225 feet (69 m). The average decline in the aquifers is 4 feet per year (1.2 m) (51). Cost problems are comparable to counties in the Trans-Pecos Region. Although analysts predict that available groundwater should remain sufficient to support continued irrigation well into the twenty-first century, pumping costs may determine future irrigation patterns in these two areas (52). Thus some experts predict that these regions will either (a) shift to growing crops (such as grain sorghum or wheat) under dryland conditions, (b) shift to less water intensive crops, or (c) move out of agricultural production altogether (53). It is not yet clear which will be the result.

A third major problem area is in the lower Rio Grande/Río Bravo reach where use of irrigation water has been heaviest in the past. Moreover, the lower reach is the area where both Texas and Mexico have continued to experience increases in demand for irrigation water. Studies of the available water on the Texas side of the Rio Grande reservoir system (based on the hydrological conditions of 1900 to 1970) indicate that water shortages will occur 70 percent of the time in the 750,000 acres (303,880 ha) awarded irrigation water rights from the Lower Rio Grande Valley Water Rights Case (54). An average annual shortage of 253,000 acre-feet (3.1 cu km) is predicted to occur in the area (55). Just to maintain existing levels of irrigation will require farmers to obtain water over and above the volumes supplied by international treaty from the Rio Grande/Río Bravo. Poor quality groundwater in the area severely limits its potential as a source of alternative irrigation water. In addition, this area is presently experiencing widespread agricultural soil-salinity problems (56). As a result, some analysts contend that the area will experience tremendous water-availability difficulties and that expansion of either agriculture or population in the lower valley may be constrained (57).

The Mexican agricultural sector has already faced water-use problems in the Lower Rio Grande/Río Bravo region and Río San Juan delta area of Tamaulipas. Difficulties were caused by an overexpansion of agricultural production, which resulted in severe drainage problems and increased salinity levels. Increased salinity levels were responsible in part for poor crop yields throughout 1960-70, in what had been "the most productive irrigated zones in the border region" (58).

Agricultural expansion in the Rio Grande/Río Bravo basin may ultimately be constrained by water availability. Thus irrigation may be limited by alternate water uses, such as municipal consumption.

Steam-Electric

Figure 5.15 graphs withdrawals of water for steam-electric cooling. These withdrawals have increased substantially between 1974 and 1980. Table 5.22 shows the dispersion of steam-electric water use in the Texas reaches. The majority of steam-electric water withdrawals in 1980 were located in the upper reach--in El Paso, Ward, and Crockett counties. Note that the middle reach had no reported steam-electric withdrawals from 1974 to 1980. Figure 5.16 shows existing electric power plants in Texas as of 1975. In 1975, this region contained a total of seven thermoelectric power plants with a total megawatt capacity of 2,153.5, with 1,045.5 megawatts being produced by steam-electric facilities.

RETURN FLOWS

Return flows provide an additional, although hard-to-quantify source of water, particularly for agricultural users. On the Mexican side, one can identify sources and destinations of return flows, but not the quantity of reuse. Texas data compiled by the TDWR suffer from several deficiencies discussed more fully below. The major discernible trend in regard to return flows is their likelihood to increase. Return flows have grown since 1978 (see Figure 5.17).

Figure 5.15

**Steam-Electric Water Withdrawals along the Texas Side of the
Rio Grande/Río Bravo Basin for 1974, 1977, and 1980**

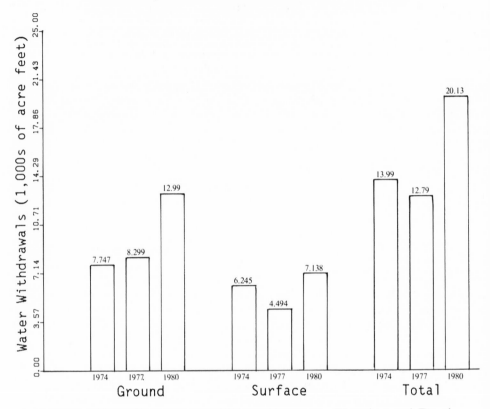

Sources: Texas Department of Water Resources, Planning and Development Division, Economics Section, "Fresh Water Use in Texas: 1974, 1977, and 1980," Austin, 1982. (Unpublished Computer Printout.)

Texas Department of Water Resources, Planning and Development Division, Economics Section, "1980 Water Use by Category and Source," Austin, 1982. (Unpublished Computer Printout.)

Figure 5.16

Existing Electric Power Generating Plants along the Texas Side of the Rio Grande/Río Bravo Basin, December 31, 1975

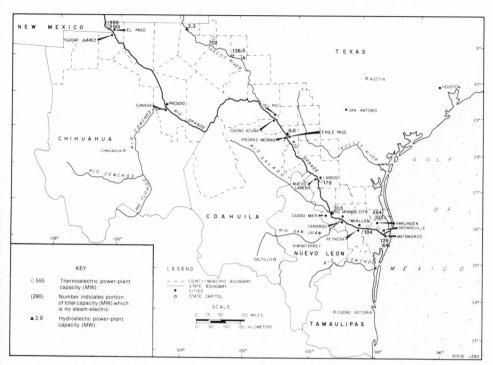

Source: Texas Water Development Board, "Continuing Water Resources Planning and Development for Texas: Phase I," Volume I, Austin, Texas, May 1977, p. II-42. (Draft.)

Figure 5.17

**Reported Municipal-Industrial Return Flows along the Texas
Side of the Rio Grande/Río Bravo Basin, 1977-1982**

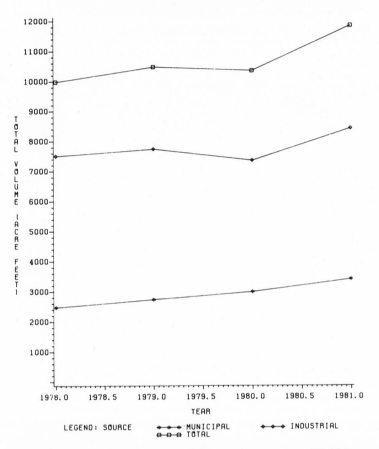

Source: Texas Department of Water Resources, "Self-Reporting System; Waste Load Data Report: 1977, 1978, 1979, 1980, and 1981,"
Austin, 1982. (Unpublished Computer Printout.)

Municipal-Industrial

The quality of TDWR's municipal-industrial return flow data is uncertain, as it reflects voluntary reporting. Each permit holder submits return-flow data based on its own samples to the TDWR. The larger the installation or facility, the more frequently TDWR requires it to sample. For example, large facilities may sample daily, while sampling in small facilities may be as infrequent as once per month (59). Such a pattern of infrequent data collection limits the reliability of quantitative estimates of return flows.

Industrial return flows have historically outpaced municipal return flows in the Texas border region, although both have increased in recent years. The majority of counties had no reported return flows. Return flows have historically been concentrated in the counties of El Paso, Cameron, and Hidalgo.

Along the Texas side of the upper reach, municipal and industrial return flows are only sporadically returned to the Rio Grande/Río Bravo. Some, especially in Crockett County, are returned to the Pecos River. In the middle and lower reaches the vast majority of return flows are discharged into a flood-diversion canal system that ultimately empties into the Gulf of Mexico (60).

Most municipal return flows in the Mexican border region are eventually returned to the Rio Grande/Río Bravo. Although some treatment may be employed prior to discharge, these return flows frequently contain large concentrations of contaminants and, in some cases, even toxic substances (61). Some reuse does occur, however. For example, some portion of Ciudad Juárez's return flow enters a canal that leads to the Valle de Juárez irrigation district. These treated residuals are used for irrigated agriculture within this district (62). Small areas of irrigation near Ojinaga and Nuevo Laredo use their return flows on an intermittent basis. The same is also true for irrigation areas bordering Piedras Negras, although in this instance the reuse occurs on a more permanent basis (63).

The region's major industrial plants have a large quantity of return flows. For example, La Domínica, S.A., discharges its return flows into a storage lake that eventually flows into La Hormiga Creek (64). Altos Hornos de Mexico, S.A., releases its return flows without prior treatment into Tornillo Creek. From there, the return flows enter the Rio Grande/Río Bravo (65).

Similarly, Refinería de Petróleos Mexicanos, S.A., releases its untreated return flows into La Escondida Lake. La Escondida Lake drains into Anhelo Creek, which feeds into the Rio Grande/Río Bravo. The polyethylene plant discharges its untreated return flows into La Rosita Basin, which eventually drains into the Anzaldous Canal. The Anzaldous Canal is also the final destination of Celanese Mexicana's return flows (66). Química Flour, S.A., treats its return flows at its installation. After treatment, these residuals are released into the Las Vacas agricultural basin, and eventually they reach Laguna Madre in the Gulf of Mexico (67). Approximately 1,772 acre-feet (0.22 million cu m) of return flows originated from these five major and three other minor industrial plants located in the Rio Grande/Río Bravo region (68).

Irrigation

Agricultural return-flow information is based on estimates of the TDWR. The TDWR does not require permits for agricultural discharges, as it considers these discharges to be nonreturnable to state waters (69).

In the upper reach, agricultural return flows have tended to be concentrated in the El Paso Valley. No usable return flows leave the immediate vicinity of El Paso or Hudspeth counties, as all return flows from the El Paso Valley are reused for irrigation in Hudspeth. In 1974, approximately 20,000 acre-feet (0.02 cu km) were reused in this area (70).

Irrigation return flows in the lower reach are concentrated in the areas with a heavy reliance on surface water--the Lower Rio Grande Valley and smaller areas in Webb, Val Verde, and Zapata counties. In 1974, about 59,000 acre-feet (0.07 cu km) of irrigation return flows originated in these areas. Of this total, about 50 percent went into the Rio Grande/Río Bravo for reuse downstream. It is not clear exactly how much water is reused for downstream irrigation (71). In fact, some analysts question the benefit of these return flows, suggesting that Texan and Mexican irrigation return flows degrade the quality of the surface water below Falcon Lake (72). Irrigation return flows in the remaining border counties tend to be negligible (73).

Mexico's border region irrigation return flows return to the Rio Grande/Río Bravo through a series of drains that lead into its tributaries. The majority of these drains are located in the lower reach. The Puertecitos, Huizache, Rancheríos, Anhelo, and Las Mujeres drains in the Bajo Río San Juan irrigation district in the state of Tamaulipas contribute approximately 4,503 acre-feet (0.55 million cu m) in annual return flows. El Morillo is the principal drain in the Bajo Río San Juan region. El Morillo discharges about 69,000 acre-feet (8.5 million cu m) annually into the Gulf of Mexico (74). Return flows in the remaining agricultural areas along the Mexican border are intermittent and of relatively inconsequential amounts (75).

An undetermined amount of these return flows are reused in the border agricultural regions. These return flows are high in salt owing to the frequency of water reuse, the high evaporation in the area, and the use of fertilizers in crop production. Additionally, the use of pesticides in the region tends to cause a further deterioration in the quality of these agricultural return flows (76).

CONCLUSIONS

Potential problem areas regarding water withdrawal exist all along the Rio Grande/Río Bravo basin. Currently there exists a situation of intense competition between sectors for available water resources. Municipal population is currently growing in the region, enhancing domestic consumption. Irrigation water use dominates at present, but its growth may be constrained by increasing municipal water demands. Consequently, current shortages may be magnified in coming years.

Are there additional sources of untapped water in the basin? Statistical evidence is, unfortunately, inconclusive. Some experts contend that the answer is negative. In fact, one recently stated that

in view of the appropriation [of surface water in the
basin], as population and economic development increase. . .
greater conservation measures will have to be taken to
stretch the available supply, and more water intensive
agricultural uses will have to be retired in favor of
municipal and industrial uses (77).

It is likely that water availability in the El Paso/Ciudad Juárez area, the Trans-Pecos Region, and specific areas in the middle and lower reaches will not be sufficient adequately to meet future water demands.

Given the historical growth rate of water withdrawal in the Rio Grande/Río Bravo basin, both Texas and Mexico will need to employ strategies aimed at resolving these existing or potential future difficulties. The City of El Paso has recently sought to employ an alternative to gain additional water--it sued New Mexico over a state law disallowing the export of groundwater across state lines. This suit has been resolved in El Paso's favor, and the city's residents may eventually obtain groundwater from nearby New Mexico aquifers. This might solve El Paso's problem, but it might also open an entirely different set of water-use problems for Mexico. Officials in Juárez are concerned that El Paso's success could lead to the eventual depletion of another aquifer which could be tapped by Mexico in order to resolve water-shortage problems (78). Conflict rather than compromise appears to be the rule in disputes over water in this area.

A similar conflict is possible in the lower reach, particularly given the expansion of population, industry, and agriculture on both sides of the border. As suggested earlier, a lack of water is projected to lead to major agricultural water shortages within this region in the coming years. Economic development plans on both sides of the border may suffer as a result.

Additional problems may be found in those regions with high groundwater-dependency rates. Rapid depletion, salinity, and especially soaring pumping costs are causing major problems in the El Paso/Ciudad Juárez area, the Trans-Pecos Region, and the Winter Garden-San Antonio area (especially in Dimmit and Zavala counties). Exploitation of the aquifers on both sides of the border suggests that Mexican farmers may be suffering from water quality and quantity problems comparable to those experienced by their northern neighbors (79).

To avoid exacerbating water conflicts, both sides may find mutual advantage in reaching bilateral agreements on water conservation. The current framework for negotiations relies upon the International Boundary and Water Commission (IBWC) as the principal forum in developing joint water agreements relating to the distribution of basin waters. In the future the IBWC may be asked to consider cooperative limits on groundwater withdrawals and joint groundwater monitoring to prevent unregulated depletion of basin aquifers. Currently those who use groundwater have little incentive to conserve, unless it costs too much to pump.

REFERENCES

1. Convention of May 21, 1906, Equitable Distribution of the Waters of the Rio Grande; United States-Mexico (34 Stat. 2953; TS 455).

2. Treaty of February 3, 1944, Utilization of Waters of the Colorado and Tijuana Rivers and of the Rio Grande; United States-Mexico (59 Stat. 1219; TS 994).

3. Interview with Norman Alford, Staff Member, Economics Section, Texas Department of Water Resources, Austin, Texas, November 12, 1982.

4. Ibid.

5. Ibid.

6. Texas Department of Water Resources, *Inventories of Irrigation in Texas: 1958, 1964, 1969, 1974, and 1979*, Report 263 (Austin: TDWR, October 1981), p. 3.

7. Ibid.

8. Interview with Alan Siles, Unit Head, Enforcement and Field Operations Unit, Shipping Control and Effluent Reports Section, Texas Department of Water Resources, Austin, Texas, January 17, 1983.

9. Comisión del Plan Nacional Hidráulico, *Esquema de Desarrollo Hidráulico para la Cuenca del Río Bravo* (Mexico, D.F.: Secretaría de Agricultura y Recursos Hidráulicos, December 1980), p. 31 (Limited Distribution Document).

10. Albert E. Utton, "An Assessment of the Management of U.S.-Mexican Water Resources: Anticipating the Year 2000," *Natural Resources Journal* 22, no. 4 (October 1982): 1097.

11. Texas Department of Water Resources, Planning and Development Division, *Water Use, Projected Water Requirements, and Related Information for the Standard Metropolitan Statistical Areas in Texas*, LP-141 (Austin: TDWR, March 1981), p. 128.

12. Stephen Paul Mumme, "The United States-Mexico Ground Water Dispute: Domestic Influence on Foreign Policy" (Ph.D. dissertation, University of Arizona, 1982), p. 153.

13. Ibid., p. 150.

14. Ibid., p. 151.

15. Ibid., p. 157.

16. Ibid.

17. Ibid.

18. Ibid.

19. Ibid.

20. Ibid.

21. *Austin American-Statesmen*, November 29, 1982, p. B-8.

22. Ibid.

23. Ibid.

24. Texas Department of Water Resources, *Water Use*, pp. 81-83.

25. Texas Department of Water Resources, Planning and Development Division, Economics Section, "Fresh Water Use in Texas: 1974 and 1977," Austin, 1982 (Unpublished Computer Printout).

26. Alford interview.

27. Texas Water Development Board, "Continuing Water Resources Planning and Development for Texas: Phase I," Volume I, Austin, Texas, May, 1977, pp. II-73, 74 (Draft).

28. Texas Department of Water Resources, "Manufacturing Water Demand: 1974 and 1980," Austin, 1982 (Unpublished Computer Printout).

29. Texas Department of Water Resources, Planning and Development Division, Water Use Technology Unit, "Reported Manufacturing Water Use: 1980," Austin, 1982 (Unpublished Computer Printout).

30. Felipe Ochoa y Asociados, S.C., *Estudio de la Calidad del Agua en la Cuenca del Río Bravo* (Mexico, D.F.: Secretaría de Agricultura y Recursos Hidráulicos, September 1978), pp. 31-38 (Limited Distribution Document).

31. Ibid., p. 33.

32. Ibid.

33. Ibid.

34. Ibid., p. 34.

35. Ibid.

36. Ibid.

37. Texas Department of Water Resources, *Inventories of Irrigation*, pp.6-7.

38. Ibid.

39. Ibid.

40. Ibid.

41. Mumme, p.157.

42. *Austin American-Statesman*, p. B-8.

43. Mumme, p. 148.

44. Ibid.

45. Neal E. Armstrong, "Anticipating Transboundary Water Needs and Issues in the Mexico-United States Border Region in the Rio Grande Basin," *Natural Resources Journal* 22, no. 4 (October 1982): 902.

46. Texas Department of Water Resources, *Inventories of Irrigation*, p.9.

47. Gordon Slogett, *Prospects for Ground-Water Irrigation: Declining Levels and Rising Energy Costs*, Agricultural Economic Report 478 (Washington, D.C.: December 1981), pp. 14-15.

48. Ibid.

49. Ibid.

50. Ibid.

51. Ibid., p.16.

52. Ibid.

53. Ibid.

54. Texas Department of Water Resources, *Water Use*, p. 128.

55. Ibid.

56. United States Water Resources Council, *The Nation's Water Resources: 1975-2000, Volume 4: Rio Grande Region* (Washington, D.C.: United States Water Resources Council, December 1978), p. 45.

57. Texas Department of Water Resources, *Water Use*, p. 126.

58. Mumme, p. 148.

59. *Austin American-Statesman*, p. B-8.

60. Ibid.

61. Felipe Ochoa y Asociados, p. 27.

62. Ibid., pp. 53-54.

63. Ibid., p. 54.

64. Ibid., p. 34.

65. Ibid., p. 33.

66. Ibid.

67. Ibid., p.38.

68. Ibid., pp. 59-61.

69. Siles interview.

70. Texas Water Development Board, p. IV-760.

71. Ibid.

72. United States Water Resources Council, p. 45.

73. Texas Department of Water Resources, "Self-Reporting System: Waste Load Data Report," Austin, 1982 (Unpublished Computer Printout).

74. Felipe Ochoa y Asociados, pp. 63-66.

75. Ibid., p. 65.

76. Ibid., p. 27.

77. Utton, p. 1097.

78. *Austin American-Statesman*, p. B-8.

79. Mumme, p.153.

CHAPTER 6

Portable
Water Systems

In developed countries most citizens have access to potable water; along the border of Mexico and the United States, this may not be the case. Expanding industrial activities, population shifts, inherent water scarcity, and deepening pockets of poverty affect the provision of potable water supplies. This chapter assesses the current drinking water supply situation along the border. The first section identifies Mexican and American institutions with jurisdiction over drinking water and outlines relevant laws, regulations, and methods of law enforcement. The second section evaluates existing water treatment facilities in the region. The third section considers technical options available for improved water treatment. The final section contains some conclusions regarding the current potable water supply situation in the region.

INSTITUTIONS AND ENFORCEMENT

A number of institutions regulate drinking water systems in the Rio Grande/Río Bravo basin. This section describes those institutions that (a) exert influence over planning, construction, and maintenance of water treatment facilities, (b) enforce drinking water quality standards, or (c) are concerned with the health ramifications of poor drinking water.

International

The International Boundary and Water Commission (IBWC), which is responsible for maintaining the boundary between the United States and Mexico, has a minor role in drinking water quality. Minute 261, signed in 1979, permits the IBWC to seek solutions to problems of "sanitary conditions that present a hazard to the health and well-being of the inhabitants of either side of the border" (1). The IBWC has yet to become involved in long-range planning for, or operation of, water systems in the Rio Grande/Río Bravo basin (2). Most planning, construction, operation, and maintenance of drinking water treatment facilities in the United States are undertaken by regional or local governments in conjunction with the federal and state governments. In Mexico, control of drinking water falls primarily in the federal government's domain. Minute 261 is important, however, because it provides a binational mechanism for water resource managers and government officials to discuss mutual health and sanitation problems.

Two other international organizations--the Pan American Health Organization (PAHO) and the United States-Mexico Border Public Health Association (AFMES)--are involved in promoting health and seeking solutions to environmental problems occurring along the border. Although neither organization enforces potability standards or controls treatment of water, each has been instrumental in enhancing public awareness of the dangers of consuming water of poor quality. Local binational health councils established by PAHO and AFMES have become valuable as a forum for discussion of joint solutions to local health problems (3).

Although these three international organizations do exist, all drinking water projects along the border are, in fact, initiated and maintained by separate national organizations in the United States and Mexico. Despite the fact that both nations have federal systems of government, drinking water treatment systems are controlled in different ways in the two nations. The United States maintains a decentralized form of government; the state and local governments play the largest role in supplying and maintaining the quality of water to citizens. Mexico, on the other hand, relies on a centralized government; control of water supply systems rests to a large degree with federal institutions. The functions of the governments of both nations, and the relevant rules and regulations of each level of government are outlined below.

United States

Prior to 1974, state governments supervised community drinking water and the federal government regulated only the quality of drinking water supplied by interstate carriers. Through the Safe Drinking Water Act of 1974 (P.L. 93-523), the federal government preempted control over most public water systems, although it now delegates oversight responsibility to the states (4).

The Act authorized the U.S. Environmental Protection Agency (EPA) to promulgate two types of national drinking water standards for primary and secondary contaminants. Primary standards specify those maximum contaminant levels (MCLs) for substances having an adverse effect on human health (see chapter 2 for a review of the drinking water standards). Interim Primary Drinking Water Standards, set by EPA in June of 1977, define MCLs or treatment techniques to achieve standards, as well as criteria and procedures for maintaining water systems (5). Secondary regulations specify MCLs protecting public welfare. These limits address issues such as the taste, odor, and appearance of drinking water. The federal government can enforce primary standards, and secondary MCLs are used only as guidelines for maintaining potable water supplies (6).

The EPA has ceded responsibility for enforcing drinking water regulations to individual states. In Texas, the Texas Department of Health (TDH) provides technical and other types of assistance to public water systems, is responsible for inspecting and monitoring drinking water sources, and it evaluates system construction and maintenance. TDH reviews and evaluates plans for new facilities, certifies water treatment system operators, and has set system performance requirements through its "Rules and Regulations for Public Water Systems" (7). TDH sets the methodology and frequency for drinking water quality testing according to the specifications of the Texas Sanitation and Health Protection Law (8).

In addition, every public drinking water system must compile a monthly water works report and submit it to TDH's Division of Water Hygiene. TDH uses these monthly reports and its own annual survey of each system to assess compliance with state and federal statutes. Those systems found violating state or federal regulations are given a grace period to correct violations. Cases of frequent or flagrant violations may be referred by TDH to either the Texas Attorney General's Office or to EPA for legal action.

Mexico

The main federal agency that regulates drinking water quality is the *Secretaría de Salubridad y Asistencia* (SSA). SSA has jurisdiction over quality control for public water systems. The *Secretaría de Agricultura y Recursos Hidráulicos* (SARH) regulates most aspects of water use; it is responsible for allocating water to various water systems.

Under President Lopez-Portillo, the government charged federal agencies with the responsibility for expanding and renovating urban potable water supply systems via the Mexican National Development Plan (9). Rural water systems are also receiving priority from the Construction and Sanitary Engineering Commission within the *Secretaría de Desarrollo Urbano y Ecología* (SDUE), which administers a self-help program for the construction of potable water supply systems (10).

In 1972, a new Subsecretariat for Environmental Improvement was established within SSA to revise and implement drinking water quality standards (11). The standards adopted for potable water were based upon those established by the World Health Organization (WHO). These standards are currently in effect, although individual Mexican states have the authority to adopt more stringent standards. State standards usually reflect attempts to control regional water quality problems in order to ensure the health and safety of residents (see chapter 2 for a listing of current Mexican drinking water standards).

SSA requires each public water system to obtain an operating permit and monitor the quality of drinking water. State SSA offices are responsible for overseeing the monitoring process. The permit and monitoring processes primarily apply to larger metropolitan areas and densely populated municipios, as many rural areas do not have formal public water systems (12).

STATUS OF EXISTING WATER TREATMENT SYSTEMS

In order to understand the problems associated with providing safe drinking water to border residents, it is important to consider the status of existing water treatment plants. This section assesses the performance of 201 regulated drinking water systems located in those Texas counties and Mexican states adjacent to the Rio Grande/Río Bravo. In addition, data are presented on the number of current residents who lack easy access to potable water.

United States

Current U.S. law defines two categories of public water systems that must adhere to federal, state, and local regulations: community and noncommunity public water systems. Other systems are not directly regulated by any level of government, although they may voluntarily maintain the same drinking water quality standards.

A *public water system* provides piped water for human consumption to either an average of twenty-five individuals for sixty days a year or at least fifteen permanent service connections. Public water systems can be classified as either *community water systems*, which serve year round residents, or as *noncommunity water systems*, which include all other public systems. Figures 6.1 and 6.2 distinguish between these two types of public systems.

According to the TDH annual surveys, there are 128 community water systems in the Texas counties bordering Mexico. In recent TDH annual surveys, about 12 percent of the community systems complied with TDH regulations, with 314 violations in noncompliant systems. The most frequent violations included improper equipment and failure to comply with TDH monthly drinking water quality testing requirements. Improper equipment violations included lack of necessary safety equipment, inadequate system design, and outdated facilities. Sixty-five percent of the systems violated these standards; the majority of these systems serve between 501 and 10,000 individuals. This high noncompliance rate may reflect the number of older municipal facilities with outdated equipment.

Approximately 34 percent of the community systems failed to comply with TDH testing requirements. Most violations were found in systems serving less than 100 individuals. A shortage of adequately trained plant operators in smaller systems may account for part of the high noncompliance rate. Other reasons include inadequate training of operators, substandard monitoring equipment, or unwillingness to self-report water quality violations.

As many of the quality violations occur in smaller water systems, a related concern is the health and safety risks that may exist in private systems too small to be regulated. In a study conducted in 1977 of the

(continued on page 138)

Figure 6.1

Community Water Treatment Systems in Texas and Mexico

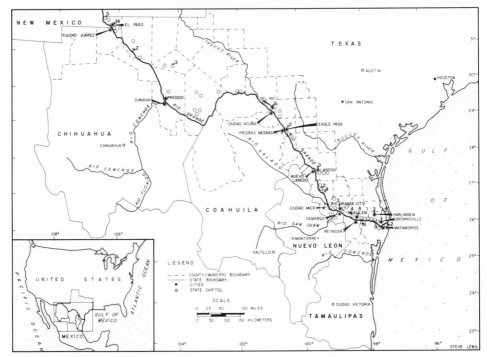

Sources: Compiled by Deborah Sagen from data obtained from 1981 and 1982 annual water system compliance surveys of the Texas Department of Health.

Felipe Ochoa y Asociados, S.C., *Estudio de la Calidad del Agua en la Cuenca del Río Bravo* (Mexico, D.F.: Secretaría de Agricultura y Recursos Hidráulicos, September 1978). (Limited Distribution Document.)

Figure 6.2

Noncommunity Water Treatment systems in Texas, 1981-1982

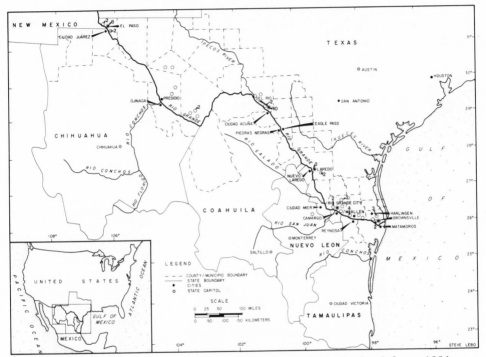

Source: Compiled by Deborah Sagen from data obtained from 1981 and 1982 annual water system compliance surveys of the Texas Department of Health.

colonias in the Lower Rio Grande Valley, the LBJ School researchers estimated that approximately 15,000 residents drew untreated water from irrigation ditches and were not served by public water systems (13). Chapter 8 on health contains a more detailed discussion of the problems related to inadequate treatment of drinking water supplies.

The problems of noncommunity water systems are similar to those of the community systems. According to the most recent TDH annual surveys, sixty-one noncommunity water systems were identified in the Texas counties studied. Many of these systems serve fewer than 100 individuals and thus receive low priority from officials. It may be difficult for TDH to correct violations because these smaller systems tend to change ownership frequently, may fail to meet with TDH officials, or may not respond to correspondence.

In these sixty-one community systems, TDH found 133 major violations. Thirteen systems, or 21 percent, were in complete compliance with TDH regulations. A large fraction of the noncommunity systems (54 percent) violated equipment or design standards. Twenty-one percent of all systems failed to comply with TDH requirements for filing monthly water quality reports.

Tables 6.1 and 6.2. list the major quality violations in the TDH files. In some instances, the addition of new equipment could significantly improve potable water quality. In other extreme cases, it might be necessary to search for a different water source.

It is hard to evaluate the health and safety risks implied by violations of TDH potable water quality standards by noncommunity systems, as these systems do not serve permanent residents. It is not possible to know the level of health risks from water provided by private systems (such as small private campgrounds or seasonal resorts) that are not required to report to TDH.

Mexico

Water is supplied through various mechanisms in Mexico; three common means are

- *toma domiciliaria* (private taps inside the residence),

- *hidrante público* (communal taps in rural areas), and

- *pipa* (purchased water delivered in trucks to street-side barrels) (14).

Most *pipa* service occurs where no formal distribution system exists (often in fast-growing cities), or where delivered water is necessary to supplement water received from an existing piped system. Project members were not able to obtain accurate or reliable information on the quantity and quality of *pipa* water in the border region.

There are data on twelve major border-area Mexican water supply systems serving cities with populations greater than 10,000. Although SARH did collect (1974) and publish (1978) a survey of those systems (15), no information was available regarding the quality of the water supply or the treatment facilities found in these cities. A majority of these systems serve less than 75 percent of the urban residents. Although some residents may rely on *pipa* service, it is possible that a number rely on water from untreated sources.

The water infrastructure problems along the Mexican side of the border are enhanced by the increase of population in this region since 1974. Although SDUE and local government agencies are attempting to increase the number of treatment facilities and update existing facilities, the financial burden is great. According to the preliminary results of the 1980 Mexican Census, it is certain that not all residents have access to adequate supplies of potable water. Chapter 8 on health provides a detailed discussion of possible health risks for persons who lack the benefit of piped potable water.

WATER TREATMENT CONTROL ALTERNATIVES

During the early years of the twentieth century, people treated water to prevent the spread of infectious disease. With the contemporary addition of urban, industrial, and nonpoint sources of pollution, the public's concern about drinking water potability is greater than ever. Water systems must be able to eliminate numerous complex contaminants in drinking water, many of which did not exist forty years ago. This section discusses some of the advanced water treatment alternatives that can improve drinking water quality in the Rio Grande/Río Bravo basin.

Water used for human consumption comes either from underground or surface sources. Groundwater is often of good quality. Treatment of groundwater may not be necessary, although disinfection is required in the United States as a precautionary measure. Some groundwater in the region contains inorganic contaminants whose removal would require sophisticated advanced treatment. Surface waters are always treated in order to remove sediments, control turbidity, and eliminate bacteriological contaminants. The various treatment techniques or *unit operations* described below are applicable to water obtained from either source. The treatment processes are presented in the order in which they might appear in a conventional surface water treatment facility. The unit operations discussed here include: aeration, coagulation-flocculation, sedimentation, filtration, ion exchange, adsorption, reverse osmosis, electrodialysis, pre- and post-chlorination (see Figure 6.3) (16).

Aeration is the process of bringing water and air into contact with each other. It achieves two purposes: removal of harmful gases from water and solution of dissolved oxygen into water.

The first step in coagulation-flocculation involves rapid mixing of a coagulant, such as aluminum sulfate, with raw water. The coagulant combines with the alkalinity in the water and forms a positively charged material that decreases the negative charges in the suspended matter. The matter forms into gelatinous masses called floc. The flocculation step involves mechanical mixing of water to enhance the formation of floc.

After floc has formed, the water is released into a sedimentation or settling tank. Sedimentation uses gravity to move water slowly and allow the settling of floc and other heavy suspended materials. The sediment (sludge) collects at the bottom of the tank. Usually sludge is removed by mechanical scrapers that force it into hoppers located at the bottom of the settling tank. Some treatment plants use two types of sedimentation to remove heavy suspended particles such as sand.

Filtration removes suspended matter from water by allowing the water to pass through a porous medium (usually sand or granular coal). Filtration alone is commonly used for treating groundwater. Filtration in conjunction with coagulation and flocculation is useful for removing suspended matter associated with treating surface water.

Water hardness is a measure of the amount of dissolved materials, usually calcium and magnesium, which exist in raw water. Two processes, lime softening and ion exchange, are used for reducing water hardness. After passing through sedimentation tanks, water is mixed with calcium hydroxide (lime) in order to remove carbonate hardness. The lime and carbonate hardness react chemically to form calcium carbonate, a precipitate that is fairly insoluble in water. Flocculation occurs next; it enhances the formation of calcium carbonate floc. The floc is then allowed to settle in a sedimentation tank where sludge is subsequently removed.

Lime softening treats noncarbonate hardness caused by salts of calcium and magnesium in addition to removing carbonate hardness. One common process adds sodium carbonate (soda ash) to the water,

Figure 6.3

Water Treatment Processes

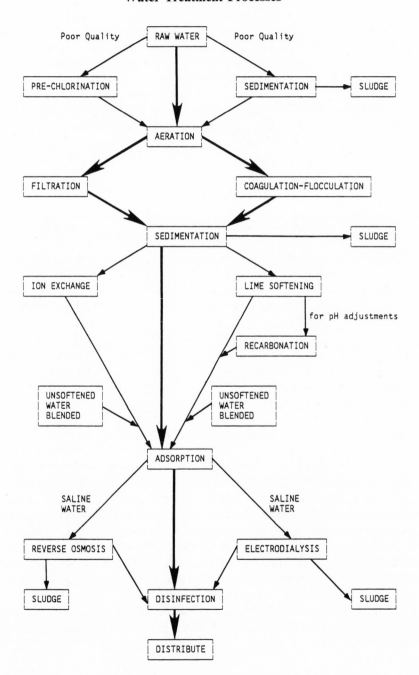

routes the water through flocculation and sedimentation, and, after treatment, adds carbon dioxide to the water to adjust the pH.

Ion exchange can also remove hardness. For example, in zeolite ion exchange, sodium ions (which do not cause hardness) are substituted for calcium and magnesium ions in the water. Water first passes through a zeolite bed, which contains large numbers of sodium ions. The sodium ions replace calcium and magnesium ions, which are then washed out of the zeolite. The zeolite bed can then be regenerated by the periodic addition of concentrated sodium chloride.

In cases where the raw water contains unacceptable levels of arsenic, fluoride, and organic pollutants, water systems may utilize an adsorption unit process. Activated carbon, for example, will adsorb chemical and organic pollutants if placed in contact with raw water. Contact tanks or powdered activated carbon are used to accomplish adsorption; the activated medium attracts pollutants that accumulate on its surface.

Some water systems must use water with a high salt concentration. Two processes that can convert saline or brackish water into a potable supply are reverse osmosis and electrodialysis. In reverse osmosis, pressure forces water from a moderately saline solution through a permeable membrane. The demineralized water then passes through the membrane, while the relatively saline water is retained and drained out of the unit. Electrodialysis applies an electric current to force saline water through special plastic membranes. The salts and minerals in the water pass through the membranes as positively and negatively charged ions. The ions are removed via the membranes and are collected as a concentrated waste stream. The demineralized water passes through the membrane, while the saline water is retained.

Disinfection is typically the last, but perhaps the most important treatment process. Disinfection destroys or deactivates pathogenic and other undesirable organisms. Chlorine, the most frequent disinfection agent, can be added to the water in gaseous or liquid form using direct or solution feeders. The chlorine then reacts with and destroys pathogenic organisms, leaving a disinfected water supply. Batch chlorine--the addition of dry chlorine directly to the water supply--is sometimes used in smaller systems, although this practice is unreliable. Batch chlorination is not recognized in Texas as an acceptable method for disinfection.

In cases where groundwater serves as the supply source, chlorination may be the only process necessary to treat water adequately. In other instances, where water quality is extremely poor, chlorination is used twice in the treatment process--once prior to any treatment (prechlorination) and again at the end of the treatment process (post-chlorination).

CONCLUSIONS

This chapter has described the institutional actors involved in regulating and operating the water treatment systems along the Texas/Mexico border. It has also discussed how these actors interact to ensure safe drinking water for border inhabitants. Although regulatory problems do exist, both nations have the technical and administrative capacity to solve them.

The performance of treatment facilities and the quality of drinking water varies greatly. In the large cities on both sides of the border the quality of drinking water is good, as there is adequate regulation, monitoring, operation, and maintenance of facilities. On the other hand, small, rural, private, or informal systems in both Texas and Mexico may be inadequately controlled. Reports on drinking water quality in small systems on the Texas side of the border indicate room for substantial improvement. Project members' inability to obtain data on the quality of small, rural, or private Mexican facilities or the water they distribute make it difficult to assess the status of these facilities.

It is also true that a few cases exist along both sides of the border where people have no access to treated water. Such residents, located primarily in poorer, rural areas or on the periphery of urban areas, risk numerous health complications by consuming water of doubtful quality. This lack of access to treated water represents a "flaw in the system." The laws of both the United States and Mexico implicitly recognize the right of all citizens to an adequate supply of potable water; solutions should be found to ensure that rural realities correspond to principled declarations.

It will not be easy to improve drinking water quality on either side of the border. How can the increasingly stringent regulations be met if there is a continuing influx of new residents? How can existing facilities adapt to a diversifying economy that generates so many novel sources of pollution? From where will come the financing for new systems or expansion of existing systems? How will the health and safety of border residents be affected if potable water is not available? Perhaps in some cases, these questions may be answered via binational mechanisms. In most cases, however, local institutions will be responsible for taking whatever actions are feasible. State and federal agencies may provide technical and financial assistance. One continuing task of water management officials will be to bring together, both along and across the border, those actors who can identify and work to resolve current and potential drinking water problems.

REFERENCES

1. International Boundary and Water Commission, "Recommendations for the Solution to Border Sanitation Problems," Minute 261 (El Paso: IBWC, U.S. Section, 1979).

2. Stephen Paul Mumme, "The Politics of Water Apportionment and Pollution Problems in United States-Mexico Relations," University of Arizona, pp. 6-7 (Undated Draft Report).

3. C. Richard Bath, "Health and Environmental Problems: The Role of the Border in El Paso-Cuidad Juárez Coordination," *Journal of Interamerican Studies and World Affairs* 24, no. 3 (August 1982): 379-381.

4. Robert M. Clark, "The Safe Drinking Water Act: Its Implications for Planning," in David Holtz and Scott Sebastian, *Municipal Water Systems: The Challenge for Urban Resource Management* (Bloomington: Indiana University Press, 1978), p. 122.

5. Ibid, p. 123.

6. Ibid.

7. "Water Hygiene," *Texas Administrative Code*, Chapter 337 (1981).

8. Ibid.

9. Stephen Paul Mumme, "The United States-Mexico Groundwater Dispute: Domestic Influence on Foreign Policy" (Ph.D. dissertation, University of Arizona, 1982), p. 151.

10. Duncan Miller, *Self-Help and Popular Participation in Rural Water Systems* (Paris, France: Organization for Economic Cooperation and Development, 1979), pp. 30-73.

11. Mumme, "Water Apportionment," p. 19.

12. Interview by Deborah Ann Sagen with Dr. Pedro Martinez-Pereda, Participating Faculty, Lyndon B. Johnson School of Public Affairs, The University of Texas at Austin, Austin, Texas, February 1983.

13. Lyndon B. Johnson School of Public Affairs, *Colonias in the Lower Rio Grande Valley of South Texas: A Summary Report*, PRP Report No. 18 (Austin: University of Texas at Austin, 1979), p. 11.

14. William J. Lloyd, "Growth of the Municipal Water System in Ciudad Juárez, Mexico," *Natural Resources Journal* 22, no. 4 (October 1982): 945.

15. Felipe Ochoa y Asociados, S.C., *Estudio de la Calidad del Agua en la Cuenca del Río Bravo* (Mexico, D.F.: Secretaría de Agricultura y Recursos Hidráulicos, September 1978) (Limited Distribution Document).

16. American Water Works Association, *Water Quality and Treatment: A Handbook of Public Water Supplies* (New York: McGraw-Hill Book Company, 1971).

CHAPTER 7

Wastewater Treatment

Control of point or nonpoint source pollution is complicated if multiple governments are involved. In the Rio Grande/Río Bravo basin, two nations plus state and local institutions are concerned with wastewater emissions. The first section of this chapter describes the relevant institutions, laws, and regulations for control of point source pollution in Texas and Mexico. The second section identifies existing municipal and industrial wastewater treatment plants. A third section discusses the technical options for control of point source pollution.

INSTITUTIONAL CONTROL

International

Although the International Boundary and Water Commission (IBWC) could in principle become involved with pollution, in practice the Commission has not taken major initiatives in this area (1,2). In 1979 the IBWC adopted Minute 261. This binational agreement gives the IBWC the authority to find solutions to "sanitary conditions that present a hazard to the health and well-being of the inhabitants of either side of the border or to impair the beneficial uses of these waters" (3).

Each of the IBWC's national sections can now take action in solving regional point source water pollution problems. The Commission has identified problems in specific geographic areas and contributed technical assistance on a case-by-case basis. As Texas and Mexico become more concerned with water quality and sanitation issues, the IBWC may seek to play a larger role in resolving related conflicts (4). The present practice, however, is to defer to the national pollution control programs of the United States and Mexico.

United States

Point source water pollution is controlled in a decentralized fashion in the United States. Although the federal government sets standards, is responsible for enforcing the laws, and appropriates funds for wastewater control, state and local governments are primarily responsible for enforcement. The Federal Water Pollution Control Act Amendments of 1972 (P.L. 92-500) are the dominant piece of water pollution law. The Act grants the Environmental Protection Agency (EPA) rule-making authority to establish stream criteria and to limit urban and industrial wastewater discharges. The EPA has subsequently established the National Pollution Discharge Elimination System (NPDES) to regulate point source discharges into navigable streams.

NPDES is a system for granting case-by-case effluent limits for wastewater dischargers, based upon general water quality criteria. The permits designate (a) specific water quality parameters for pollutants, (b) individual specifications for monitoring/reporting of effluent quality, and (c) compliance deadlines for each discharger.

In Texas, the EPA Region 6 office cooperates with the Texas Department of Water Resources (TDWR) in administering the NPDES program. TDWR's Wastewater Section prepares draft NPDES permits which become legal upon EPA approval. Compliance and enforcement operations are handled by both the Enforcement and Field Operations-Municipal Wastewater and Water Use Section, and the Industrial Wastewater and Solid Waste Sections of TDWR (5). In addition to the NPDES program, TDWR operates urban and industrial permit systems for two categories of wastewater facilities, those which allow effluent to be discharged, and those which require retention of effluent. TDWR permits for these plants include:

● limitations for chemical and biological constituent emissions in domestically treated wastewater,

● criteria specifying conditions for reuse of effluent in irrigation, and

● requirements for temporary discharge of untreated or partially treated wastewaters (6).

Individual permits may impose strict limitations on the quality and quantity of discharged effluents. Each wastewater facility is responsible for complying with the terms of its permit and submitting results of quality monitoring on a regular basis.

Thus, the TDWR issues permits for all facilities within the state, under both national and state systems. In some cases facilities may be required to obtain both NPDES and TDWR permits, a situation conducive to bureaucratic overlap. This system of self-reported compliance monitoring is neither easily accessible nor completely accurate. Although the TDWR receives thousands of effluent reports each month, these may not always be reliable, as wastewater treatment plant operators may lack sufficient training to perform the sometimes complicated quality testing. Monitoring may also be affected by the inherent contradiction in a self-reporting system; an operator has few incentives to admit to flagrant permit violations (7).

TDWR uses a quarterly compliance summary report to list dischargers who fail either to submit reports or to meet permit requirements for the quarter. TDWR also conducts periodic reviews of all permits once every five years. An extensive computerized information management system is used to compile most monitoring information (8). This tracking mechanism is a definite asset, although it does not alleviate a lag time between water quality violations and TDWR action.

The Federal Water Pollution Control Act Amendments also require water quality management planning. Section 208 of the Act addresses area-wide planning for urban, industrial, and nonpoint source water pollution problems. Area-wide management plans are developed by local agencies in designated areas, or by the states in areas without regional planning agencies. Section 208 designated areas are those in which a particular part of a river has been identified as having a substantial water quality control problem due to point source pollution (9). For example, the Lower Rio Grande Basin region has been designated as a 208 planning area.

An important component of the 208 planning process is the identification of projected public wastewater treatment facility needs. Such planning reports are compiled by a variety of local and regional agencies along the U.S. side of the border, although the Lower Rio Grande Valley Development Council is the only agency with direct 208 planning responsibilities. Once completed, these reports are certified by the

Governor of Texas as part of the Texas Water Quality Management Plan. Plans provide information on existing facilities, projected public facilities, their cost implications, and alternative proposals for meeting treatment needs. Plans currently exist for 1983, 1990, and 2000.

The federal and state mechanisms for control of point source discharges allow municipalities in the border region to develop their own schedules for the planning, construction, and maintenance of wastewater treatment facilities. Communities most often operate their facilities as public utilities and use bonding authority and user fees to cover their costs. With increasing populations and cuts in state and federal funding, border cities have been placed in a difficult situation. If they build, they step more heavily into debt and must increase user fees. If they try to keep costs down, the plans to update wastewater treatment facilities may slip further down on a lengthening list of priorities.

Mexico

The control of point source water pollution is centralized at the federal level in Mexico. Federal agencies, most notably the *Secretaría de Agricultura y Recursos Hidráulicos* (SARH), are responsible for controlling point source water pollution. State offices are assigned only a minor cooperative role in control and enforcement activities. Local governments do manage and construct wastewater treatment facilities.

A 1971 amendment to the Mexican Constitution, the *Ley Federal para Prevenir y Controlar la Contaminación Ambiental*, granted SARH authority to regulate both the quality of discharged effluents and construction of municipal wastewater treatment facilities (10).

In complying with this amendment, SARH enacted a set of regulations entitled *Reglamento para la Prevención y Control de la Contaminación del Agua*. These regulations require all public or private establishments to (a) register the wastewater emissions and characteristics with SARH, (b) comply with effluent standards, and (c) present plans for the control of settleable solids, oil, grease, temperature, and pH. In addition, SARH coordinates registration of point discharges with the *Secretaría de Salubridad y Asistencia* (SSA). A particular discharger may choose either to treat effluents jointly with other industries and municipalities (through water quality control districts) or provide their own treatment that complies with a SARH discharge permit. Dischargers are given three years after registration with SARH to comply with the terms of their permit.

In practice, however, SARH, SSA, and other branches of the Mexican government have encountered difficulties in enforcing these regulations. All levels of Mexican government--federal, state, and local--face heavy financial burdens. The National Urban Development Plan, established in 1978, has helped to alleviate some of the financial burden. This Plan targets federal monies for the improvement of sanitation and sewerage facilities to border cities (11). Funding can also be sought from international agencies, as Juárez is attempting to do for expansion of its wastewater facilities (12).

A second factor affecting enforcement is the shortage of skilled personnel within SARH and SSA to monitor facilities, administer the permit program, and provide technical assistance to public and private dischargers. Border municipalities also face problems in hiring and training skilled plant operators (13).

A third factor is that the centralized system of Mexican water pollution control forces border cities to compete with other regions for both funds and attention. Local governments must receive federal approval for any action to solve local sewer and sanitation problems.

WASTEWATER TREATMENT FACILITIES

The first step in this assessment of the current status of wastewater treatment in the Rio Grande/Río Bravo basin is to identify existing municipal and industrial wastewater treatment facilities. For the purpose of this chapter, the Rio Grande/Río Bravo basin is defined to include the Texas counties and the Mexican states which are directly adjacent to the river.

Municipal and industrial wastewater facilities are easily identified in Texas because the TDWR permit system is, in principle, an open record available from the TDWR computer system. TDWR keeps information on plant location, permit levels of biochemical oxygen demand, total suspended solids, and dissolved oxygen. Permit records also list allowed as well as self-reported effluent flow. Municipal plant permits also record the type of treatment. Field investigations by project members found that TDWR records are not always complete and that the TDWR cannot easily gain access to data on actual effluent quality (14).

Information on the municipal and industrial wastewater treatment plants in Mexico has been obtained from a study conducted for SARH in 1978 (15). Although SARH requires the registration of those who discharge wastes, project members were unable to obtain data on the current status of municipal and industrial facilities. The 1978 SARH study does include data on the actual quality of effluent, the point of discharge, types of treatment used, and percentage of population served.

Municipal Wastewater Facilities

Forty-four municipal wastewater treatment facilities have been identified in thirteen border counties of Texas. Of these forty-four facilities, twenty-nine hold TDWR permits to discharge effluent into the Rio Grande/Río Bravo or its tributaries. The remaining fifteen facilities do not release effluent to surface waters. One major problem identified in the records is that the volume of inflows commonly exceeds plant capacity during the summer months when water use increases. TDWR records do not indicate that any of the border water treatment plants constitute a major water pollution or sanitation problem.

Project members were able to obtain information on twelve municipal wastewater treatment facilities along the Mexican side of the Rio Grande/Río Bravo. Of these facilities, six discharge effluent for irrigation. Seven discharge directly to the Rio Grande/Río Bravo, its tributaries, or connected irrigation drains. One discharges to the Gulf of Mexico, one utilizes an oxidation pond, and five provide no treatment. Nontreatment can be a problem in larger cities such as Juárez, where the volume and the quality of the effluent can significantly affect surface water quality (16).

Figure 7.1 maps the locations of these fifty-six municipal wastewater facilities identified on both sides of the border. Facilities that discharge effluent to surface waters are denoted with an open circle; those plants that utilize oxidation ponds are indicated by triangles. The facilities are for the most part located in large urban areas.

Because of the difficulties of locating private, unregulated sewerage systems, it is assumed that areas which have no municipal facilities are served by septic tanks or other private waste disposal systems. This assumption may not be reasonable, as selected 1970 and 1980 Mexican and U.S. census data indicate serious problems with lack of or inadequate plumbing and sewerage facilities.

Table 7.1 lists 1970 census data on sewage and plumbing facilities for residences in selected Texas counties. These numbers indicate that between 1.9 (El Paso) and 19.5 (Hidalgo) percent of all homes are not served by any formal sewerage facilities. Perhaps these homes have outdoor pit privies or substandard indoor plumbing facilities.

Figure 7.1

Municipal Wastewater Treatment Systems: Texas and Mexico

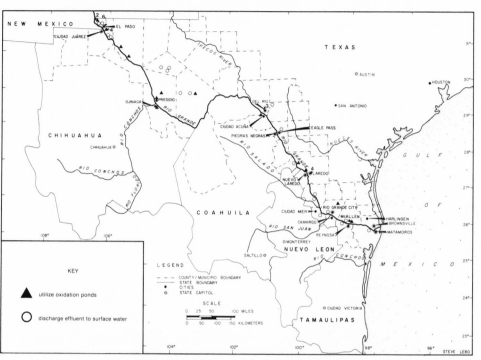

Sources: Felipe Ochoa y Asociados, S.C., *Estudio de la Calidad del Agua en la Cuenca del Río Bravo* (Mexico, D.F.: Secretaría de Agricultura y Recursos Hidráulicos. September 1978). (Limited Distribution Document.)

Texas Department of Water Resources, "DW2525 Report Facility Types," Austin, 1982. (Unpublished Computer Printout.)

Table 7.2 lists comparable 1970 census data for selected Mexican cities. Mexican water distribution systems are categorized by the location of the water: none, inside, or outside of the home. Here again there are a number of homes (as many as 25.6 percent of all homes in Acuña, for instance) which do not have access to plumbing or sewerage facilities.

Table 7.3 gives an indication of the percentage of Mexican homes, both urban and rural, which lack sewerage facilities. These preliminary 1980 Mexican census data show that approximately 43 percent of all homes in all three states lack public or private sewerage systems. Comparing Tables 7.2 and 7.3 reveals that rural areas may have a greater problem with lack of adequate sewerage facilities than major urban areas.

Industrial Wastewater Facilities

Industries in the Rio Grande/Río Bravo basin either dispose of their wastewater via municipal treatment plants or operate their own facilities. Fifty-four industrial wastewater treatment plants operate in nine of the fourteen Texas border counties. Although descriptions of the industrial treatment processes are listed on TDWR permit applications, the TDWR does not computerize such information. Figure 7.2 maps all of the industrial facilities in Texas. Facilities denoted with a triangle indicate plants which retain their effluent; an open circle indicates plants which discharge effluent.

A 1978 SARH study listed wastewater treatment facilities for eight major industries along the Texas/Mexican border. Of these eight industries, four provide no treatment, two provide primary treatment, and two do not describe their treatment processes. Four facilities discharge effluent into the Rio Grande/Río Bravo or its tributaries; two discharge for irrigation (17).

Because there are only sixty-two industries with private wastewater treatment facilities, one may assume that industrial water pollution is not a problem in the region. This may not be true, as the tracking mechanisms for industrial facilities are not as effective as they are for municipal facilities. For example, TDWR does not include treatment processes in its computerized information system because it is difficult to classify these systems unambiguously. Moreover, it is not known if those industries using municipal sewerage systems are placing a strain on existing municipal facilities. Industrial residuals can be different and

Figure 7.2

Industrial Wastewater Treatment Facilities: Texas and Mexico

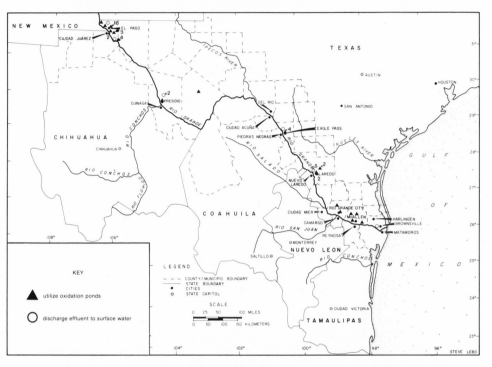

Sources: Felipe Ochoa y Asociados, S.C. *Estudio de la Calidad del Agua en la Cuenca del Río Bravo* (Mexico, D.F.: Secretaría de Agricultura y Recursos Hidráulicos, September 1978). (Limited Distribution Document.)

Texas Department of Water Resources, "DW2525 Report Facility Types," Austin, 1982. (Unpublished Computer Printout.)

more difficult than municipal effluent. Industrial wastes may require advanced water treatment processes in order to ensure good water quality.

WASTEWATER TREATMENT CONTROL ALTERNATIVES

What types of physical, chemical, and biological processes can be used to treat the wastewater along the Rio Grande/Río Bravo, considering the sources of municipal and industrial waste? This section describes one set of treatment alternatives chosen because they are applicable to the Rio Grande/Río Bravo basin. This list is, of course, not intended to be all-inclusive of the options available for treating wastewater.

Private residences in rural areas, schools, parks, and facilities which release under 20,000 gallons of effluent per day often rely on septic systems. Septic tanks collect effluent and allow bacterial action to treat raw wastewater. Solids are removed and oxygen demanding materials are reduced by biological treatment via anaerobic digestion. The effluent is then distributed via an underground pipe into a leaching field where it percolates into the ground.

More advanced treatment processes must be used for larger sewerage systems, especially in urban areas and for industrial wastes. A typical wastewater facility would provide preliminary, primary, and secondary treatment. Advanced or tertiary treatment can be used for treating complex effluents or waste streams which contain high levels of contaminants. Figure 7.3 summarizes these processes.

Preliminary and primary treatment processes are used to remove large, coarse solids. Physical unit processes are used most often, although chemical treatment can be utilized. Four types of primary treatment are well-suited for use in the Rio Grande/Río Bravo basin: wastewater pumping, preliminary treatment, conventional primary sedimentation, and chemically induced primary sedimentation.

Secondary treatment processes are used in conjunction with primary treatment in order further to reduce organic constituents of raw wastewater. After primary treatment has removed coarse solids, biological or secondary treatment can be used to remove organic substances that are either soluble or of colloidal size. Different types of secondary treatment include: activated sludge and its variations (extended aeration, contact stabilization, and high rate activated sludge), low and high rate trickling filters, rotating biological contactors,

Figure 7.3

Wastewater Treatment Processes

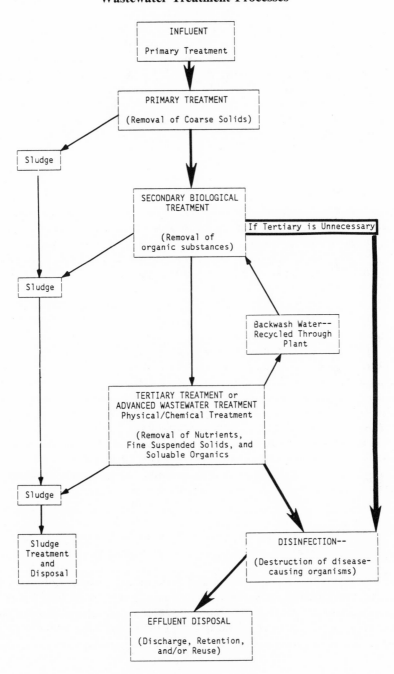

oxidation ditches, and aerated or oxidation ponds. Conventional and chemically induced secondary sedimentation is used with those processes that generate large volumes of sludge, as it removes many of the microorganisms used during secondary treatment.

Advanced wastewater treatment, or the use of physical and/or chemical tertiary treatment, may be necessary if secondary treatment is insufficient to meet effluent quality standards. Physical/chemical tertiary treatment can remove nutrients (such as nitrogen and/or phosphorous), suspended solids and soluble organics (including toxic pollutants). Generally a settled secondary effluent is used as tertiary influent. Tertiary treatment processes include selective ion exchange, biological nitrification/denitrification, multimedia and sand filtration, carbon adsorption, tertiary two-stage lime treatment, and ammonia stripping.

Disinfection, which is usually the last step in the treatment process, can also be used to pretreat raw wastewater of extremely poor quality. Disinfection is most often accomplished through the use of chemical agents such as chlorine gas, chlorine dioxide, hypochlorous acid, or ozone. Physical agents, mechanical means, or ultraviolet radiation are rarely used, as they are more costly alternatives. Disinfection selectively destroys disease-causing organisms, such as bacteria which may cause typhoid, cholera, paratyphoid, bacillary dysentery, and other water related diseases.

Sludges are the solids removed from bulk liquid wastewater during the course of treatment. Sludges contain much of the objectionable pollutants from wastewater and must be treated and disposed of properly. The most common form of sludge treatment and disposal used in rural areas is a combination of digestion or stabilization, dewatering, and land disposal. Other unit processes to be discussed include pumping, thickening, conditioning, incineration, landfilling, and land application of sludge, aerobic digestion, anaerobic digestion, lime stabilization, wet oxidation, and heat drying.

The disposal of effluent occurs after raw wastewater has passed through all stages of treatment. Effluent can be disposed either by returning it to a natural water source, to land, or to some other user--a municipality, an industry, or for agricultural or recreational uses. Requirements for effluent quality usually determine the method of disposal. Disposal methods most applicable to the Rio Grande/Río Bravo basin include: discharge to surface waters, infiltration-percolation, overland flow, subsurface injection, reuse, and recycling.

In planning wastewater treatment facilities, many factors must be taken into account. The amount and type of wastewater to be treated must be considered, particularly for industrial wastewaters which may require special forms of treatment. The location of the plant may also affect the type of treatment and disposal methods. Government regulations can affect alternatives by restricting the physical or biological composition of sludges. Finally, environmental impacts must be assessed so the discharged effluent and sludges will not disrupt the natural environments to which they are returned. In essence, the type of control alternative selected will depend on numerous and often conflicting factors.

CONCLUSIONS

Control of point source water pollution is an important aspect of water resources management. One method for reducing water pollution is to regulate municipal and industrial wastewater treatment facilities. In the Rio Grande/Río Bravo basin, both the U.S. and Mexican governments have control programs for planning, constructing, and operating wastewater treatment facilities. The U.S. water pollution control program issues permits for waste dischargers, enforces wastewater treatment regulations, and encourages regional planning and management of wastewater. In Mexico the central government controls the construction of new municipal wastewater treatment facilities and regulates discharges. Both nations need either to improve treatment plants or to build new ones as the population in sister cities along the border increases. Both nations have problems financing improved wastewater treatment and hiring trained personnel.

A problem unique to the region is that the Rio Grande/Río Bravo is an international boundary. Point source pollution cannot be confined to national borders, so that effective control requires an element of international cooperation. As point source water pollution becomes a larger problem in the Rio Grande/Río Bravo basin, more comprehensive binational solutions may be sought. Wastewater treatment facilities in the region can be improved, as could the enforcement programs of both nations. Although it is becoming more expensive to treat wastewater, technical solutions do exist if there is the political will to implement them. If both nations commit themselves and continue to adjust their management strategies, improve technology, and keep the lines of communication open within and between the two nations, point source water pollution can be effectively controlled.

REFERENCES

1. C. Richard Bath, "Health and Environmental Problems; The Role of the Border in El Paso-Ciudad Juárez Coordination," *Journal of Inter-American Studies and World Affairs* 24, no. 3 (August 1982): 383.

2. Stephen Paul Mumme, "The Politics of Water Apportionment and Pollution Problems in United States-Mexico Relations," University of Arizona, pp. 18-20 (Undated Draft Report).

3. International Boundary and Water Commission, United States Section Minute 261, "Recommendations for the Solution to the Border Sanitation Problems," September 24, 1979.

4. Mumme, pp. 7-8.

5. Texas Department of Water Resources, *Report of the Texas Department of Water Resources for the Biennium: January 1979-August 1981* (Austin: TDWR, 1981), pp. 28, 30.

6. *Texas Water Code*, Sections 5.131 and 5.132, "1979 Amendments to Effluent Standards," Section 156.18.05.001-004, and "General Regulations Incorporated Into Permits," Section 156.18.05.001-010.

7. Telephone Interview with Sandra Johnson, Permit Control Reports Section, Texas Department of Water Resources, Austin, Texas, October 19, 1982.

8. Lyndon B. Johnson School of Public Affairs, 1982 Policy Research Project, "Water Quality in the Rio Grande River: An Environmental and Institutional Assessment," Austin, July 1982, p. 31 (Draft).

9.Comptroller General of the United States, *Better Data Collection and Planning Is Needed to Justify Advanced Waste Treatment Construction* (Washington, D.C.: General Accounting Office, December 1976), p. 2.

10. *Ley Federal para Prevenir y Controlar la Contaminación Ambiental*, Chapter 3, Articles 14-22, 1971.

11. Mumme, p. 21.

12. Bath, p. 384.

13. Interview with Dr. Pedro Martinez-Pereda, Participating Faculty, Lyndon B. Johnson School of Public Affairs, The University of Texas at Austin, Austin, Texas, February 4, 1983.

14. Data compiled by Deborah Sagen from the Texas Department of Water Resources files during 1982-1983.

15. Felipe Ochoa y Asociados, S.C., *Estudio de la Calidad del Agua en la Cuenca del Río Bravo* (Mexico, D.F.: Secretaría de Agricultura y Recursos Hidráulicos, September 1978) (Limited Distribution Document).

16. Ibid.

17. Ibid.

CHAPTER 8
Water-Related Diseases Along the Texas/Mexico Border

The relationship between water and health has been recognized for centuries. Moses and the Prophet Mohammed formulated laws and rules for sanitation and personal hygiene to protect the health of their followers. Relics of old water supplies and sewage systems uncovered in ancient Babylonia, Egypt, Athens, and Rome are evidence of the importance that these cultures attached to the preservation of water for community use (1). However, scientific understanding of sanitary practices is relatively new. Only in 1855 did Snow infer a relation between cholera and contaminated water; later Budd showed that typhoid spread via water supplies (2). At present, twenty to thirty infective diseases are associated with contaminated water supplies (3).

The purpose of this chapter is to explore the nexus between water, wastewater, and water-related diseases along the Texas-Mexico border. The first section describes the major classifications of water-related diseases: waterborne, water-washed, water-based, and water-related vector disease. The second section discusses the water infrastructure along the border which can affect the incidence of such diseases. A final section presents recent morbidity and mortality indicators of the principal enteric diseases affecting Texas counties and Mexican municipios along the border. These statistics appear to indicate that several water-related diseases are endemic in the Mexican municipios and largely eliminated in the Texas counties. These conclusions are necessarily tentative as substantial uncertainties exist regarding the accuracy and reliability of disease reports, particularly in Mexico. Reported cases of enteric disease in the Mexican municipios vary considerably from one year to the next or among contiguous municipios. The diversity of health systems in Mexico also complicates the complete reporting of information.

CLASSIFICATION OF WATER-RELATED DISEASES

Water-related diseases traditionally are classified by the causal microbe as viral, bacterial, protozoal, or helminthic. A second classification that may be more helpful for considering the relation of diseases to improved water supplies was proposed by David J. Bradley in 1968 (4). This classification rests upon four distinct mechanisms of transmission. Table 8.1 lists the most important water-related diseases and one estimate of the percentage of reduction that can result from water improvements. For example, guinea worm can be completely eliminated. The incidence of other diseases, such as yellow fever and infective hepatitis, may be little altered by water improvements. Table 8.2 describes both mechanisms and corresponding prevention strategies.

Waterborne diseases are contracted through the drinking of contaminated water which acts as the passive carrier for pathogenic organisms. Cholera, typhoid, infectious hepatitis, and giardiasis are important waterborne diseases that fall in this category. Feachem has argued that the term *waterborne* disease is greatly abused by public health or water engineers who apply it indiscriminately so that it has become almost synonymous with water-related disease (5). He points out that a source of misunderstanding has been the assumption that if a disease is waterborne, then water must be the only means of transmission. In fact, all waterborne diseases can be transmitted by any route that permits fecal material to be ingested. For instance, cholera may be contracted via contaminated food. One strategy to reduce waterborne diseases is to improve drinking water quality.

A *water-washed disease* is one that decreases as the volume of water used for hygienic purposes is increased, regardless of the quality of the water (6). Three main types of water-washed diseases can be differentiated. One type includes infections of the intestinal tract, such as diarrhoeal diseases. These diseases are important causes of serious morbidity and mortality, especially among infants in tropical climates. Feachem argues that diarrhoeal diseases, particularly shigellosis, decrease with the availability and volume of water used. These diseases may be strongly associated with the microbiological quality of the water (7). A second type of water-washed disease includes infections of skin and eyes resulting from a poor personal hygiene. Scabies and trachoma are diseases in this category. The third type consists of infections carried by fleas, lice, mites or ticks, which can be reduced by improving personal hygiene. One important prevention strategy for all of these water-washed diseases is to improve water quality and personal hygiene.

Most *water-based diseases* are infections by parasitic worms, such as schistosomiasis and guinea worm, which depend on aquatic organisms to complete their life cycle. Aquatic snails serve as intermediate hosts for infective cercarie that are the cause of schistosomiasis. The schistosome larvae develop until infective cercariae are shed into the water and reinfect man through his skin. The transmission mechanism of guinea worm is similar, in that the guinea worm escapes from humans through skin lesions and develops in small aquatic crustacea. Infection occurs when such water is ingested. One preventive strategy is to control the intermediate host population.

Some *water-related diseases* are spread by insects that breed in water or bite near water. Malaria, yellow fever, dengue, and onchocerciasis are transmitted by insects that breed in water. Trypanosomiasis (sleeping sickness) is transmitted by the tsetse fly, which bites near water. Hence, one important preventive strategy is to destroy breeding sites of insects.

Other diseases that analysts have related to the quality of the water include goiter, food poisoning, metallic poisoning, fluorosis of the teeth, and nonspecific intestinal derangements (8). Although the addition of sodium iodide to water may help prevent goiter, sodium iodide in common table salt has proven a more reliable method. Lead poisoning may occur when ingestion exceeds 0.3 to 0.6 milligram per day per capita. Although small quantities of copper, zinc, and iron may be beneficial to health, large concentrations may derange bodily metabolism sufficiently to cause disease. Water with high concentrations of fluorides may produce fluorosis. However, a minimal amount of fluorides is desirable in potable water to aid in the prevention of dental caries. Therefore the addition of fluorides to water, known as fluoridation, is widely practiced.

DETERMINANTS OF DISEASE

A number of factors related to water consumption have been recognized as important determinants of morbidity and mortality due to intestinal infectious diseases (9). These include the

● availability of drinking water,

● quality of water supplies,

● distribution of sanitary facilities, and

● housing characteristics.

It is almost axiomatic that an improvement in these factors can lead to an improvement of the public health (10). One principal source of water contamination is sewage, which contains many pathogenic microorganisms that cause disease in man. Other means of contamination include leakage of polluted water into wells or other sources of underground water, floods, relaxation of vigilance in the purification of a public water supply, droppings by birds of carrion into reservoirs, or the creation of breeding places for mosquitoes and other flying insects (11).

The availability of piped water and sewage disposal in the homes of people are fundamental components in any plan for eradication of water-related diseases (12). Tables 8.3, 8.4 and Figure 8.1 show the availability of piped water for six Texas counties and four Mexican states in 1980. Hidalgo, Maverick, and Cameron present the highest percentage of persons in housing units lacking complete plumbing (see Table 8.3). In Hidalgo 9.9 percent of all housing units lack complete plumbing; these homes represent 11.6 percent of the population. Comparable housing-units figures for Maverick are 12.4 percent and 11.2 percent, respectively, and for Cameron 8.3 percent and 9.7 percent, respectively. Val Verde has the lowest fraction of people without piped water, 2.4 percent.

Comparable 1980 census data for the Mexican border municipios have not yet been published. Table 8.4 lists the closest equivalent water facilities in the Mexican states of Chihuahua, Coahuila, Nuevo León, and Tamaulipas in 1980. The contrast between U.S. border counties and Mexican states in water infrastructure is dramatic. In the three states an average of 20 percent of all housing units lack complete plumbing and 45 percent lack public drainage. Although significant improvements have been made in the supply of piped water in the Mexican states, much can still be accomplished in the implementation of sewage systems. Because sewage is one of the principal sources of water contamination, reduction or eradication of waterborne diseases will be limited as long as many Mexican housing units remain without sewage systems. If the problem of waste disposal were solved in the Mexican states, the high rates of morbidity and mortality due to enteric diseases would be reduced.

Figure 8.1

Water Facilities for Selected Texas Counties and Mexican States, 1980

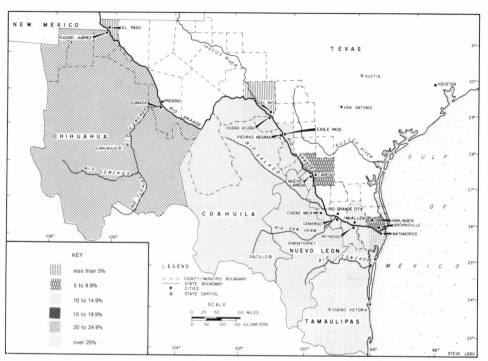

Source: Calculations by Alfonso Ortiz Nunez based on U.S. Department of Commerce, Bureau of Census, *Characteristics of Housing Units*, chapter A, vol. 1, part 45 (Washington, D.C. U.S. Government Printing Office, 1982).

MORBIDITY AND MORTALITY DATA

This section presents the incidence rates of selected water-related diseases for some Texas and Mexico border counties and municipios. U.S and Mexico data are discussed separately, although the rates in each country will later be compared. U.S. data are reliable to the extent that the computerized system of vital statistics of the Texas Department of Health contains the relevant information. According to many people interviewed the reliability of the Mexican information is limited because not all communicable diseases are reported. Since several different health systems operate in Mexico, conflicting channels of communication and authority appear to exist. There are in fact three parallel health systems: Servicio de Asistencia Social, Instituto Mexicano de los Seguros Sociales, and Instituto de Salubridad y Seguros Sociales para los Trabajadores del Estado. This diversification impedes accurate compilation of health statistics.

Tables 8.5 through 8.11 present the most recent morbidity and mortality data for enteric disease in Texas border counties and Mexico border states available in published sources. There are still some gaps for certain years and diseases. Table 8.5 lists reported rates of selected enteric diseases per 100,000 population for some key Texas border cities. Salmonellosis and shigellosis present the highest rates, ranging to 37.2 and 63.5 in Laredo and Brownsville, respectively. Amebiasis, typhoid fever, and infectious hepatitis have low and perhaps insignificant rates for all localities except for Laredo, which still has high rates in comparison with Texas and the United States (see Table 8.7).

Table 8.6 appears to indicate that counties have lower rates than their major cities. Salmonellosis and shigellosis are the leading causes of reported enteric diseases in border counties. The shigellosis rate in Cameron, 33.4, is two times the rate of shigellosis for Texas and four times the rate for the United States (see Table 8.7). The salmonellosis rate for Webb County, 55.4, is high when compared with levels of 17.3 and 14.9 for Texas and the United States, respectively. It is interesting to note the low rates of reported cases for amebiasis and typhoid fever in all the counties. Near-zero rates for these diseases in Texas and the United States in 1980 are listed in Table 8.7. Infectious hepatitis rates are below endemic indices. Although cases reported from 1976 to 1979 appear to indicate a rising rate of enteric disease, this apparent trend may be explained in part by substantial improvement in the reporting system of communicable diseases in most of the local health departments.

Table 8.8 shows reported enteric disease rates in 1976-79 for the four Mexican states bordering Texas: Tamaulipas, Coahuila,

Chihuahua, and Nuevo León. The pattern of water-related diseases in the Mexican states is different from Texas, as enteric diseases are a major health problem in all the Mexican states (see Figure 8.2). The rates of reported cases of enteric diseases appears to reflect an endemic and pervasive character of these diseases in Tamaulipas, Coahuila, Chihuahua, and Nuevo León. For example, amebiasis in the state of Tamaulipas increased from a rate of 55.0 cases per 100,000 in 1976 to 703.4 in 1979. This tremendous rate of increase may not reflect a real increase but rather more accurate reporting of detected cases. The rate of amebiasis for 1978, 802.5, is more in accord with levels reported for areas of similar development and sanitary conditions; for example, Coahuila and Chihuahua (1979) and Nuevo León (1978) have reported amebiasis rates of 1283.9, 572.2, and 616.2 cases per 100,000 population, respectively. These rates are up to 500 times greater than those of Texas and the United States.

Reported salmonellosis rates are disconcerting. Coahuila presents a rate of 241.2 in 1978, while neighboring states of Tamaulipas and Nuevo León have a rate of only 3.5 per 100,000 population. Chihuahua registered a reported rate of 67.6. These data illustrate the unreliability of reporting on communicable diseases in Mexico.

Typhoid rates are higher than those for Texas. In 1976, Coahuila registered a rate of 24.9 cases of typhoid fever compared with a rate of 0.5 for Texas. Shigellosis occurs at a rate moderately above the pattern in Texas. One possible inference from the data in Table 8.8 is that the pattern of enteric diseases in the Mexican states bordering Texas may be cyclical and unstable, reflecting epidemic outbreaks of these diseases.

Tables 8.9 through 8.11 list mortality rates due to enteric diseases for selected Texas counties and Mexican municipios. Tables 8.9 and 8.10 indicate that mortality due to enteric diseases has been practically eradicated in the six Texas counties. From rates of 8.2 cases per 100,000 population in 1970, the six counties passed to a rate of zero or almost zero in 1981. The enteric-disease mortality rates in Mexican municipios are significantly higher (see Table 8.11). These data, from 1978, are the most recent information available. Enteritis shows an epidemic character; it accounts for a substantial percentage of the deaths of children in the age group of 0-4 years. Ciudad Juárez, Piedras Negras, and Reynosa have rates of death ranging from 60.3 to 35.4 per 100,000 population, respectively. Such rates are similar to those that exist in the rest of Mexico or in other developing countries, as indicated by Pan American Health Organization studies in infant mortality in Latin America (13). A current study of infant mortality in the city of Nuevo Laredo shows that gastrointestinal diseases are responsible for 19.5 percent of all the deaths occurring to children under one year of age in 1981 (14). Although there may be many causes of such high mortality, the predominant differences between the Mexican and U.S. side of the border

Figure 8.2

**Reported Rates of Amebiasis for Selected Texas Counties, 1980 and
Mexican States, 1979
(number of cases per 100,000 population)**

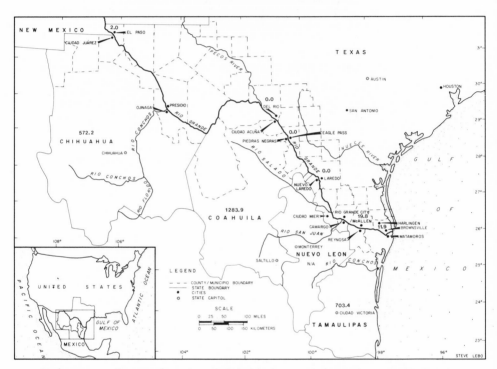

Sources: Secretaría de Salubridad y Asistencia, Unidad de
Información, "Reporte de Enfermedades Comunicables, Mexico 1976-
1979," Mexico, D.F., 1979. (Unpublished Data.)

Texas Department of Health, Bureau of Communicable Disease Services, *Reported Morbidity and Mortality* (Austin: Division of Public
Health Education, 1981).

N/A = not applicable.

are poorer infrastructure development, sanitary conditions, housing, and the general standard of living.

CONCLUSIONS

Water-related diseases remain a major problem in the Mexican municipios bordering Texas. Such diseases appear to continue to occur with endemic and epidemic character. The high mortality and morbidity rates may be related in part to the large fraction of people living in houses without piped water and sewerage systems. On the other hand, the counties in Texas bordering Mexico have practically eradicated the incidence of such diseases; this pattern may reflect the number of people having plumbing and sewage disposal systems. Such improvements in the distribution of water, housing facilities, and public education are fundamental factors in the eradication of water-related diseases. The task of providing safe drinking water supplies and sewage disposal systems is one of enormous magnitude, particularly in developing countries like Mexico. The economic dilemma is very difficult. To prevent the occurrence and spread of water-related diseases, the United States and Mexico may seek to place special emphasis on the protection of the quality of surface and underground waters and early detection of waterborne disease outbreaks.

REFERENCES

1. Gabre E. Teka, *Water Supply--Ethiopia* (Addis Ababa: University Press, 1977), p. 1.

2. David J. Bradley, "Health Aspects of Water Supplies in Tropical Counties," in *Water, Wastes and Health in Hot Climates*, ed. Richard Feachem, Michael McGarry, and Duncan Mara (New York: John Wiley & Sons, 1977), p. 5.

3. Ibid., p. 6.

4. Richard Feachem, "Infectious Disease Related to Water Supply and Excreta Disposal Facilities," *Ambio* 6, no.1 (October 1977): 59.

5. David Bradley, "Health Problems on Water Management," *Journal of Tropical Medicine and Hygiene* 73, no. 1 (June 1970): 286-293.

6. Feachem, p. 58

7. Ibid.

8. Gerald Berg, *Transmission of Viruses by the Water Route* (New York: Interscience Publishers, 1965), p. 5.

9. Ruth Puffer and Carlos Serrano, *Mortality in Childhood* (Washington, D.C.: Pan American Health Organization, 1973), p. 309.

10. Harold E. Babbit and James Doland, *Water Supply Engineering*, 5th ed. (New York: McGraw-Hill Book Co., 1955), p. 366.

11. William Sarles, *Microbiology: General and Applied* (New York : Harper & Brothers, 1951), p. 379.

12. Puffer and Serrano, p. 373.

13. Ibid.

14. Alfonso Ortiz Nuñez, "Comparative Study of Infant Mortality in the Texas-Mexico Border Area of Laredo/Nuevo Laredo" (M.P.A. Professional Report, LBJ School of Public Affairs, University of Texas at Austin, 1983).

CHAPTER 9

Forecasting
Water Use

The patterns of water withdrawals and water pollution indicate that water users along both sides of the Rio Grande/Río Bravo face increasing water scarcity and reduced water quality. As a result, public officials and water planners must seek to develop alternative means of providing adequate water resources despite the complications imposed by the relative poverty of the region and aridity of the natural environment. Water management must in turn be based upon a reliable assessment of the likely volume of water that various users will demand in the future.

One purpose of this chapter is to present a set of forecasts for water use in thirty-three Texas counties in the Rio Grande/Río Bravo basin. A second purpose is to demystify water-use forecasts by highlighting the procedures and assumptions of applied forecasting methodologies. This section defines some basic concepts. A subsequent section describes various projection techniques. These techniques are illustrated by the Texas Department of Water Resource's (TDWR's) forecast of water demands for various sectors on the Texas side of the Rio Grande/Río Bravo.

Two common terms for discussing future water uses are *water demands* and *water requirements*. These terms have different meanings but are often used interchangeably. In cases where the price of water or income of user is not considered to affect the quantity of water withdrawn, the terms *water needs* or *water requirements* are used. *Water demand* refers to the amount of water that would be requested at various prices, assuming that increased costs lead to decreased demands. It is difficult to calibrate the exact volume of demands at various prices because water users are not always responsive to price changes and prices for some users rarely change. Studies on the price and income elasticities of water use have found that a user's income does influence per capita water use (1). Other studies have found that residential water use is relatively unaffected by price (2). The fact that water demands may be inelastic to price changes does tend to undermine the use of price for forecasting future water withdrawals.

FORECASTING METHODS

Forecasting models may be classified by level of disaggregation, the time horizon, purpose, and type of mathematical model. Forecasts can be made on an aggregated or disaggregated basis. Sectoral forecasts break water use into classes, so that future water use can be related to variables specific to a given class. For example, municipal water-use forecasts may rely on population projections, industrial use is based on estimates of output, and agricultural estimates are tied to irrigated acres of land. Sectors can then be aggregated to a total withdrawal figure for a municipality, region, state, or nation.

Time horizons can range from short term (for example, a day-to-day analysis of an industrial or municipal water system) to long term (such as a fifty-year projection assessing whether a region will be in a water deficit position at some time in the future). Day-to-day forecasts may permit a manager of an existing water system to adjust system flows to changes in weather or fluctuations in river flows. Short-term forecasts could also be used to determine the quantity of water needed by a manufacturing plant to respond to changes in product demand. Short-run models typically include many variables. Longer horizons (such as twenty-five years) are used by municipalities to project the size and technological characteristics needed for future municipal water and wastewater systems. A typical long-term forecast is highly aggregated and subjective; it would include only the few variables considered critical for projecting future water use. Value judgments and subjective analysis become more crucial as the time horizons extend in length.

A first step in developing a forecast model is to define the relevant water-use sectors. Chapter 5 considered historical water withdrawals for the manufacturing, municipal, steam-electric, mining, irrigation, and livestock sectors. This chapter projects water withdrawals for municipal, steam-electric, irrigation, and livestock sectors, as well as an industrial sector (comprised of a combined mining and manufacturing sector).

Four strategies for forecasting water demands/requirements include the use of *time extrapolation, a single coefficient requirement model, a multiple coefficient requirement model,* and *a multiple coefficient demand model.* These four alternatives can be used to create a wide range of applications. One extreme is the immediate forecast, with many endogenous variables and a few fixed exogenous variables. The other extreme is the long-range model, with one or two variables and many fixed exogenous parameters. The choice of variables, model, and level of aggregation is often influenced by the availability of information, time, and funding.

Perhaps the simplest projection is an extrapolation of a rate of change observed over some time period. In this method a variable is selected which has been correlated with water use in the past. It is assumed that the variable will continue to reflect water use in the future. This method does not consider the causes of the increase and overlooks factors which may affect the rate of growth. There is no *a priori* or behavioral theory of water use underlying the extrapolation. For example, water use per capita may have increased by 2 percent per year during the recent past. An extrapolation would extend that rate into the future without considering the available supply of water or other constraining factors.

A single coefficient model forecasts water use by projecting one or more critical variables thought to be related to the observed increase. The forecaster must assume that water use bears a fixed relationship to the chosen variable. For instance, one could project per capita water use by projecting the number of water-using appliances, such as washing machines and dishwashers. This method requires an *a priori* theoretical relationship between some independent variable and water use. Unlike time extrapolation, which links water-use changes to the continuation of a recent trend, the single coefficient method ties water use to the increase or decrease of another event.

Multiple coefficient *requirement* models and multiple coefficient *demand* approaches use regression techniques to correlate past variations in water use with numerous variables. These models are similar to the single coefficient method, in that they require a theory on which to construct the model. Requirement and demand models differ in the use of income or price as variables. A water demand model includes variables that reflect the theoretical interaction between income or price and the volume of water used to perform some function. That is, as price increases the volume of water used should decrease. A water requirement model assumes that price and income do not significantly affect water use, and these factors are omitted from the model.

TEXAS FORECASTING METHODS FOR THE RIO GRANDE/RÍO BRAVO BASIN

The Legislature of the State of Texas requires the Texas Department of Water Resources (TDWR) to forecast Texas water requirements fifty years into the future. This section outlines the techniques that TDWR uses to project water demands/requirements for the industrial, steam-electric, irrigation, livestock, and municipal sectors. TDWR uses a different projection method for each sector; each requires a substantial volume of historical and projected data and a theoretical basis on which to construct a mathematical relationship. The following sections dissect the industrial, agricultural, and municipal sectors forecasts. Each section describes the appropriate data base, reproduces the projection equations, and shows how calculations of future water demands/requirements are derived. Forecasts will be reported by reach (upper, middle, and lower) and irrigation area (Trans Pecos, Winter Garden, and Rio Grande Valley). The counties which compose those areas are listed in Table 9.1.

Industrial Projections

Figure 9.1 illustrates the industrial projection process. TDWR projects water withdrawals for all two-digit Standard Industrial Classification (SIC) industries using a single coefficient model. Water use is considered to be a function of projected industrial growth, defined in terms of value added through production. Five SIC sectors, accounting for 90 percent of industrial water use in Texas, are considered major water users: processed foods and beverages (SIC 20), paper (SIC 26), chemicals (SIC 28), petroleum refining (SIC 29), and primary metals (SIC 33). For these sectors TDWR reviewed estimates of future growth rates for two-digit SIC industries published by public agencies, private sector economists, and independent industry experts, as well as its own study. The agency chose the U.S. Bureau of Economic Analysis's (BEA's) fifty-year projections of earnings and employment (based on data provided by the U.S. Bureau of Labor Statistics) as a basis to calculate water withdrawals for the chemicals, primary metals, food

Figure 9.1

TDWR Procedure for Forecasting Industrial Water Withdrawals

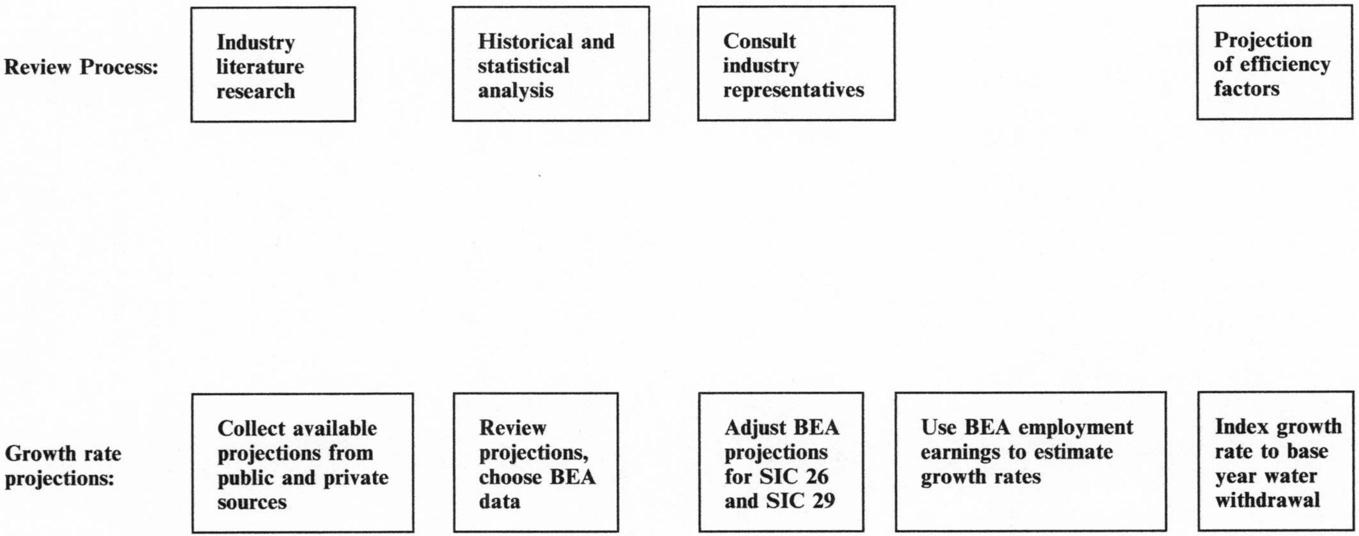

Review Process:

| Industry literature research | Historical and statistical analysis | Consult industry representatives | Projection of efficiency factors |

Growth rate projections:

| Collect available projections from public and private sources | Review projections, choose BEA data | Adjust BEA projections for SIC 26 and SIC 29 | Use BEA employment earnings to estimate growth rates | Index growth rate to base year water withdrawal |

Source: Texas Department of Water Resources, Planning and Development Division, Water Requirements and Uses Section, "Texas Industrial Water Use Long Term Projections," Austin, 1982, p. 5. (Draft.)

processing, and all minor water-using industries. TDWR independently projected petroleum refining and paper sector growth rates (4). It also developed compound annual growth rates for all two-digit SIC industrial sectors, assessed these rates for reasonableness, and indexed them to base year water use. To arrive at future year water demands/requirements, TDWR then multiplied these growth rates (as adjusted for efficiency factors) by base year water use.

Each TDWR industry growth rate is based on decade-to-decade changes in Gross Product Originating (GPO). To determine the GPO, TDWR uses BEA projections of Gross National Product (GNP), income projections for each state two-digit industry, and income projections for the corresponding national two-digit industry (see Equation 9.1 illustrated in Figure 9.2). A compound annual growth rate (CAGR) is calibrated from differences in projected GPO over the base year GPO. For example, Figure 9.3 presents the formula and computation of the 1980-90 CAGR for the chemicals industry. This growth rate is then indexed to base year water use, so that future year water use is a function of the CAGR and projected efficiency factors (see Figure 9.4 for the 1990 chemical industry projection). Table 9.2 lists high and low industrial water projections for the upper, middle, and lower reaches along the Rio Grande/Río Bravo.

Steam-electric water withdrawal projections are based on assumptions concerning the type of future generating capacity and type of facility--coal, lignite, natural gas, fuel oil, or nuclear fuel. Data on existing and planned (to the year 2000) electrical power plants are derived from 1982 publications of the Electric Reliability Council of Texas, the Southwest Power Pool, and the Western Systems Council (5). TDWR estimates that total generating capacity beyond 2000 will be proportional to increases in the population and growth in each two-digit industrial sector. TDWR estimates for the steam-electric sector are shown in Table 9.3.

Agricultural Projections

TDWR uses a linear program to project irrigation water demands (see Figure 9.5). The modified formulation maximizes profits, subject to mass balance constraints on allowable acreage for each individual crop and the amount of potentially irrigable land (Figure 9.6). A demand schedule can be derived by varying the price of water and listing the quantities of water used at each level of water cost. Figure 9.7

(continued on page 184)

Figure 9.2

**TDWR Equation for Estimating Future Gross
Product Originating (GPO)**

$$GPO(i,t,TX) = GNP(t) * \frac{Y(i,TX)}{Y(i,US)} \quad \text{for all } i, t \quad (Eq. 9.1)$$

where

GPO(i,t,TX) = gross product originating for the ith industry in
               in year t, expressed in constant dollars

GNP(t) = gross national product in year t adjusted by
          removing SIC 26 and SIC 29

Y(i,TX) = income projections for the ith industry in Texas

Y(i,US) = national income projections for the ith industry

t = an index of industry SIC

i = an index of year.

Source: Texas Department of Water Resources, Planning and Development Division, Water Requirements and Uses Section, "Texas Industrial Water Use Long-Term Projections," Austin, 1982, p. 9. (Draft.)

Figure 9.3

Derivation of 1980-1990 Growth Rate for Chemicals Industry
(GNP and GPO expressed in billions of constant 1972 dollars)

Step 1:   **1980 Estimate of GPO**

$$GPO(1980,TX) \quad = \quad GNP(1980) \; * \; \frac{Y(1980,TX)}{Y(1980,US)} \qquad (Eq.\ 9.2)$$

$$2.1743978 \quad = \quad (1,459.7) \; * \; \frac{1.353}{908.28556}$$

Step 2:   **1990 Estimate of GPO**

$$GPO(1990,TX) \quad = \quad GNP(1990) \; * \; \frac{Y(1990,TX)}{Y(1990,US)} \qquad (Eq.\ 9.3)$$

$$3.5222131 \quad = \quad 2022.1 \; * \; \frac{2.357}{1353.152}$$

Step 3:   **Absolute Growth**

1980 to 1990   =   GPO(1990,TX) - GPO(1980,TX)

1.347732   =   3.52213 - 2.1743978

Step 4:   **Growth Rate Per Year**

1.347732   =   4.941348 percent increase per year

Source: U.S. Department of Commerce, Bureau of Economic Analysis, "Long-term Regional Projections, Income, Employment and Population," Washington, D.C., 1981. (Computer Tape.)

Figure 9.4

Equation for Forecasting Industrial Water Withdrawals

$$IWD(t) = IWD(b) \, (XGR)^{(t-b)} \, (RxT) \qquad (Eq.\ 9.4)$$

where

IWD(t)   = Projected industrial water demand at time t

IWD(b)   = Base year industrial water demand

$XGR^{(t-b)}$ = Indexed growth rate from base year to projection year

RxT   = Conservation in water use from improved technology and practices

Calculation of 1990 Water Withdrawals for Chemicals Industry
(thousand acre feet)

$$IWD(1990) = IWD\,(1980)\,(XGR)^{(t-b)}\,(RxT) \qquad (Eq.\ 9.5)$$

$$882.6 = (558.1)\,(1.049)^{10}\,(.977)$$

Source: Texas Department of Water Resources, Planning and Development Division, Water Requirements and Uses Section, "Texas Industrial Water Use Long-Term Projections," Austin, 1982, p. 64. (Draft.)

Figure 9.5
TDWR Procedure for Forecasting Irrigation Water Withdrawals

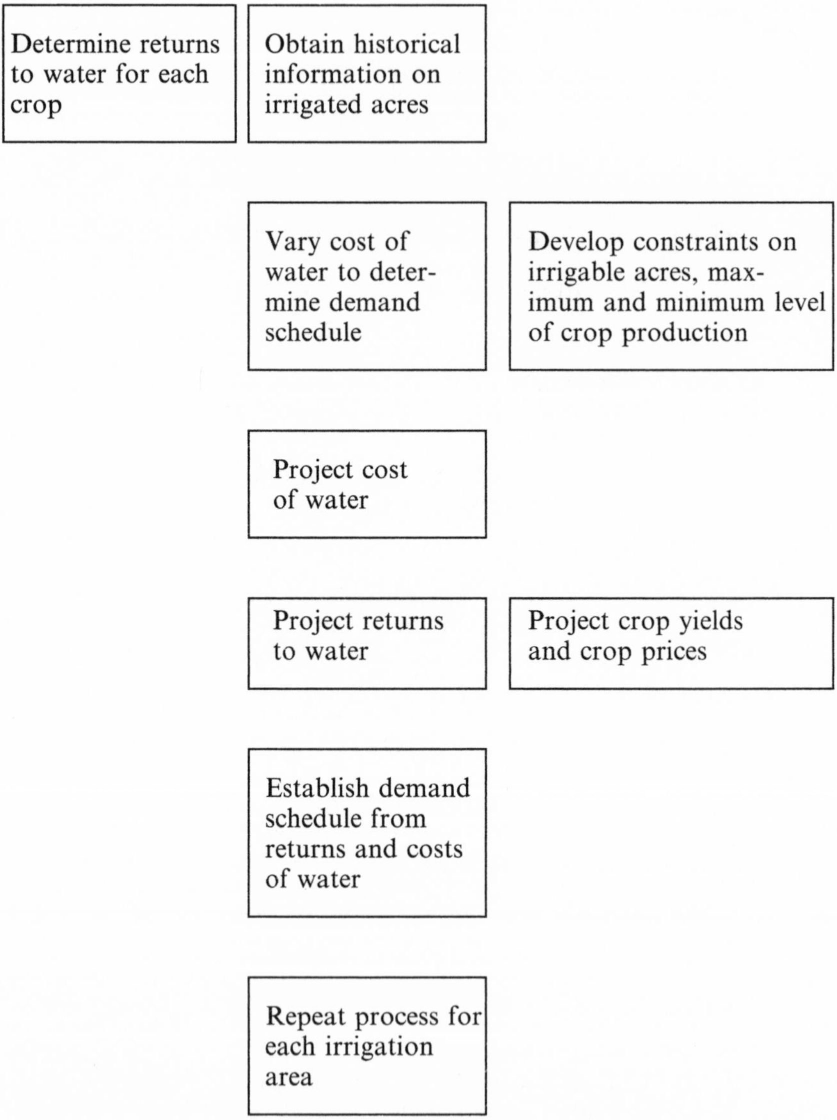

Determine returns to water for each crop	Obtain historical information on irrigated acres

Vary cost of water to determine demand schedule	Develop constraints on irrigable acres, maximum and minimum level of crop production

Project cost of water

Project returns to water	Project crop yields and crop prices

Establish demand schedule from returns and costs of water

Repeat process for each irrigation area

Source: Texas Department of Water Resources, Planning and Development Division, Water Requirements and Uses Unit, "Projecting Future Food and Fiber Water Requirements for Texas," Austin, 1982. (Draft.)

Figure 9.6

Linear Programming Formulation

Maximize $Z = \sum\limits_{i=1}^{n} C(i)D(i)$ (Eq. 9.6)

subject to

$$\sum\limits_{i=1}^{n} C(i) \leq M$$

D(i) - K(i) = f(i) - (p * w(i))

C(i) ≤ m(i)

n(i) ≤ C(i)

where

K(i)	=	negative profit for crop i
D(i)	=	positive profit for crop i
C(i)	=	number of irrigated acres of crop i planted
f(i)	=	dollar of profit per irrigated acre
p	=	dollar price of water per irrigated acre per year
w(i)	=	water volume required to produce crop i on an irrigated acre
M	=	maximum number of acres for all crops that can be irrigated
m(i)	=	maximum number of acres of crop i that can be irrigated
n(i)	=	minimum number of acres of crop i that can be irrigated.

Source: Texas Department of Water Resources, Planning and Development Division, Economics Section, Water Requirements and Uses Unit, "H2ODEM," Austin, 1982. (Computer Tape.)

illustrates the process for calculating water demands. Irrigation water demands are determined by the water requirements of each crop and the number of irrigated acres at a given price of water.

For the purpose of forecasting future water demands, projections of returns to water can be compared with projections of water costs to determine if irrigation would be profitable for each crop. The method used to determine returns to water, the cost of supplying water, and methods for projecting each of these factors are discussed below.

To determine the profits gained by irrigation, returns to all factors of production except water were obtained for a base year (1980) from crop enterprise budgets from the Texas Agricultural Extension Service (6). The residual profit was assumed to be the maximum amount farmers would pay for water. Thus, if the projected cost of water is greater than the projected residual, irrigation would not be profitable.

The resource restrictions include a limit on potentially irrigable acreage which is defined as total acreage used for all crops (dryland and irrigated land) in 1980. Upper and lower bounds on the acres plantable for each crop reflect the upper and lower historical levels (expressed in percentage of total crop acreage) multiplied by potentially irrigable acres.

Crop yields and future commodity prices are two external factors used to project future returns to water. Increased crop yield is a function of technological advances and improved seed varieties. Rates of water application per acre are reduced over time for each area, a practice which implicitly assumes improved irrigation. Commodity prices are projected by the U.S. Department of Agriculture's National Inter-Regional Agricultural Projections (NIRAP) model. This simulation model of the domestic food and agricultural system uses regional U.S. crop production and world market conditions to estimate national market-clearing commodity prices. Price estimates for commodities not included in NIRAP were obtained either from private sources or were linked to the prices of other closely related commodities.

Future water costs also reflect prices of electricity and natural gas for well and pump operations. Three price scenarios (low, medium, and high cost of electricity and gas) are used to estimate a range of water costs.

Livestock water projections are generated from data on the number of animals and the per animal water requirements. The total head of livestock is based on projections of U.S. food and fiber requirements, upper limitations on land available for grazing, and the assumption that Texas maintains its historical share of U.S. livestock production. Table 9.4 lists border-area forecasts of irrigation and livestock water withdrawals, by irrigation area.

Figure 9.7

**Calculation of 1990 Water Withdrawals for
Sample Irrigation Area**

$$IWD = \sum_{i=1}^{n} C(i)w(i) \qquad (\text{Eq. } 9.7)$$

where

C(i)   = number of irrigated acres of crop i planted

w(i)   = water volume required to produce crop i
                   on an irrigated acre

IWD   = total irrigation water demands

If the price per acre foot of water is below the break-even point for
alfalfa, the model maximizes irrigated acres of alfalfa and minimizes
irrigated acres of cotton. Irrigation water demand would be:

Alfalfa   (147,423 acres) (2.76 feet/acre) =   406,887.5

Cotton   (528,830 acres) (.58 feet/acres) =   306,721.4

Total for High Plains District 5 =   713,608.9

11 the price per acre-foot of water exceeds the break-even point for
alfalfa, the model minimizes irrigated acres of alfalfa and maximizes
irrigated acres of cotton. Irrigation water demand would be:

Alfalfa   (16,906) (2.76 ft/acre) =   4,660.56

Cotton   (659,347) (.58 ft/acre) =   3,421.26

Total for High Plains District 5 =   8,081.82

Source: Texas Department of Water Resources, Planning and Development Division, Economics Section, Water Requirements and Uses
Unit, "H2ODEM," Austin, 1982. (Computer Tape.)

Municipal Projections

Municipal water withdrawal projections are based on changes in population and per capita use (see Figure 9.8 for an overview of the projection process). A multiple regression model is used to relate per capita water use with various independent variables, including total annual precipitation, total annual pan evaporation, total median family income, and price of water per thousand gallons.

TDWR runs a regression of historical water use on those factors for each city over 1,000 population, by year. Figure 9.9 shows the regression model and Table 9.5 gives the results obtained by using a statewide data set. Given a projection of each independent variable, TDWR uses the regression equation to project municipal water use for various years; the per capita change in water use reflects the differences between the base year and each successive projection year. Table 9.6 lists TDWR's projections of municipal water withdrawals by reach for the period 1980 through 2030.

CONCLUSIONS

Table 9.7 aggregates irrigation, municipal, industrial, and livestock forecasts to project future water withdrawals by each use sector and reach. Projected water-use patterns for irrigation and livestock sectors are related to three major irrigation areas. Changes in projected municipal and industrial water use are described for the lower, middle, and upper reaches. All of these forecasts are based on the previously discussed TDWR projection methods.

Irrigation water withdrawals were 88 percent of total basin water use in 1980. The "high projections" (see Table 9.7) indicate that the share of all water used by irrigation is expected to decline from 81 to 67 percent over the fifty-year projection period. Offsetting this decline

Figure 9.8

TDWR Procedure for Forecasting Municipal Water Withdrawals

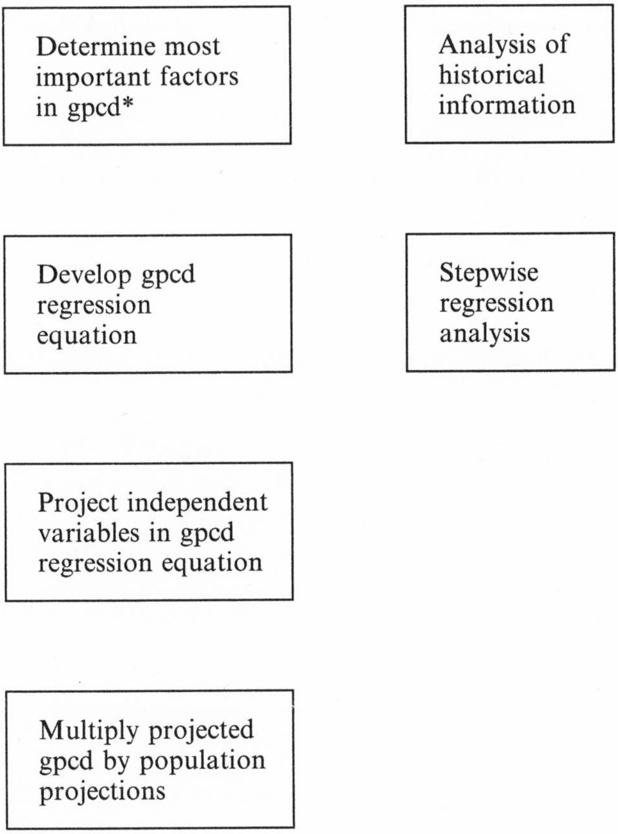

Determine most important factors in gpcd*	Analysis of historical information
Develop gpcd regression equation	Stepwise regression analysis
Project independent variables in gpcd regression equation	
Multiply projected gpcd by population projections	

Source: Texas Water Development Board, Economics Division, "Methodology for Projecting Municipal Water Needs," Austin, 1976. (Draft.)

* gallons-per-capita-per-day.

Figure 9.9

TDWR Regression Equation for Estimating Per-Capita-Per-Day Water Withdrawals

$$Y = a + b(1)logX(1) + b(2)logX(2) + b(3)logX(3) + b(4)logX(4)$$

where

Y	=	per-capita water use in gallons per day
X(1)	=	total annual precipitation in inches
X(2)	=	total annual pan evaporation in inches
X(3)	=	total annual median family income
X(4)	=	per thousand gallon price of water

Source: Texas Department of Water Resources, Planning and Development Division, Economics Section, Water Requirements and Uses Unit, "Methodology for Projecting Municipal Water Needs," Austin, 1976, p. 23. (Draft.)

is an increase in the share of municipal water use; it increases from 10 percent in 1980 to 31 percent of total basin withdrawals by 2030. TDWR projects both industrial and livestock use at approximately 1 percent of basin withdrawals.

Irrigation water use in 1980 was roughly distributed as follows: 45 percent for the lower Rio Grande area, 26 percent for the Winter Garden area, and 29 percent for the Trans-Pecos area. TDWR projects a substantial shift to occur in irrigation withdrawal patterns by 1990. The Trans-Pecos irrigation area is projected to increase its portion of withdrawals from 29 to 47 percent. The Winter Garden area is projected to withdraw less water relative to the other areas in the basin, with its fraction of use dropping from 26 to 11 percent; the lower Rio Grande area is projected to drop from 45 to 42 percent.

In 1980, Winter Gardens withdrew the largest fraction of water for livestock purposes, approximately 61 percent. The lower Rio Grande uses 13 percent and the Trans-Pecos accounted for 26 percent of livestock withdrawals. Livestock withdrawals are expected to rise until the year 2000, when a stable demand and constraints on grazing land may limit sector growth.

With respect to total basin municipal water withdrawals, the lower reach accounted for 44 percent, the middle for 11, and the upper for 45 percent. In 1990 these proportions are expected to change to the following percentages: 48 for the lower, 10 for the middle, and 42 for the upper reach.

Hildalgo, Cameron, and Webb counties withdrew the largest amount of municipal water in the lower reach. Hildalgo accounted for 39 percent, Cameron for 32 percent, and Webb for 19 percent. In 1990, Hidalgo and Cameron are expected to raise municipal use in the lower reach from 39 to 41 percent, and from 32 to 35 percent, respectively. Webb is projected to decrease its proportion from 19 to 16 percent.

In the middle reach, Val Verde, Uvalde, and Maverick counties were the three largest users in 1980, accounting for 38, 20, and 17 percent, respectively. In 1990, Maverick is expected to overtake Uvalde as the second largest municipal water withdrawer. Maverick is projected to withdraw 24 percent, Uvalde 19 percent, and Val Verde 35 percent of municipal water in the middle reach. In the upper reach, El Paso is the major municipal water user, using about 81 percent of the total municipal water withdrawals. The next largest user is Reeves County with 3 percent. In 1990 El Paso is expected to use 86 percent of the reach's municipal water.

Industrial water withdrawal patterns for 1980 showed that the upper reach used 60 percent of all the water used for the industrial sector. The lower reach used 32 and the middle 7 percent. Industrial water use in the lower reach is dominated by Cameron and Hidalgo counties. These two counties used 94 percent of the industrial water in that reach in 1980, and the 1990 fractions are expected to remain at the same level. In 1980, Zavala County withdrew 87 percent of the water for industrial purposes in the middle reach. By 1990, this fraction is projected to decline to 86 percent. In the upper reach, El Paso uses 98 percent of industrial-use water and is projected to stay at this level. Overall, the upper reach used 60 percent of total basin industrial water.

REFERENCES

1. Roger H. Willsie and Harry L. Pratt, "Water Use Relationships and Projection Corresponding with Regional Growth, Seattle Region," *Water Resource Bulletin* 10, no. 2 (April 1974): 361.

2. Robert C. Camp, "The Inelastic Demand for Residential Water: New Findings," *American Water Works Association* 10, no. 8 (August 1978): 453.

3. Freilich and Litner, Consultants, *Water and Wastewater Master Plan, City of Austin, Existing Wastewater Facilities* (Boston: Metcalf and Eddy, Inc., 1982), p. 15.

4. Texas Department of Water Resources, Planning and Development Division, Economics Section, Water Requirements and Uses Unit, "Texas Industrial Water Use Long-Term Projections," Austin, 1982, p. 7 (Draft).

5. Texas Department of Water Resources, Planning and Development Division, Economics Section, Water Requirements and Uses Unit, "Water Planning Projections for Texas--1980-2030," Austin, 1982, p. 21 (Draft).

6. Texas Department of Water Resources, Planning and Development Division, Economics Section, Water Requirements and Uses Unit, "Projecting Future Food and Fiber Water Requirements for Texas," Austin, 1982, p. 4 (Draft).

Reference
Materials

— **TABLES**

— **GLOSSARY**

— **BIBLIOGRAPHY**

Table 1.1

Physical Characteristics of the Rio Grande/Río Bravo Basin

Section of River	Length (miles)	Surface Area (square miles)		Total Surface Area (square miles)
		Mexico	U.S.	
Origin to El Paso/Ciudad Juarez	745	-0-	4,742	4,742
El Paso/Ciudad Juarez to the confluence of the Rio Conchos	290	29,240	34,089	63,339
Confluence of the Rio Conchos to Falcon Dam	684	43,523	48,919	92,442
Falcon Dam to the Gulf of Mexico	274	15,961	1,208	17,169

Source: Felipe Ochoa y Asociados, S.C., *Estudio de la Calidad del Agua en la Cuenca del Río Bravo* (Mexico, D.F.: Secretaría de Agricultura y Recursos Hidráulicos, September 1978), p. 8. (Limited Distribution Document.)

Table 1.2

Multi-Year Average Annual Climatic Extremes Recorded in the Rio Grande/Río Bravo Basin

Category	High Value	Low Value
Precipitation	24 inches (63.5 cm)	10 inches (25.4 cm)
Evaporation	111 inches (282 cm)	78 inches (198 cm)
Windspeed	5.4 mph	3.1 mph
Relative Humidity	75%	62%
Temperature	74 degrees F (23 C)	68 degrees F (20 C)

Source: International Boundary and Water Commission of the United States and Mexico--United States Section, *Flow of the Rio Grande and Related Data, 1980*, Water Bulletin 50 (El Paso: IBWC, 1980), p. 7-78.

Table 1.3
Wind Movement in the Rio Grande/Río Bravo Basin
(miles per hour)

Segment	Jan	Feb	Mar	Apr	May	Jun	Jul	Aug	Sep	Oct	Nov	Dec	Annual
Station 1: Martin King Ranch, Texas													
1980 Ave.	3.7	4.6	6.0	6.0	6.6	8.4	7.2	7.1	5.2	3.9	3.6	3.3	5.5
1957-1980 Ave.	3.9	4.6	6.1	6.2	6.7	7.1	6.7	5.9	5.0	4.6	4.0	3.5	5.4
Station 2: Amistad Dam, Texas													
1980 Ave.	3.0	3.5	4.2	4.2	3.6	4.5	4.0	4.0	3.4	2.9	2.8	2.8	3.6
1963-1980 Ave.	3.3	3.7	4.5	4.6	4.5	4.8	4.5	4.0	3.5	3.4	3.1	3.1	3.9
Station 3: Eagle Pass, Texas													
1980 Ave.	3.8	4.5	5.5	5.4	5.6	6.2	5.9	5.7	4.7	3.7	3.3	3.5	4.8
1963-1980 Ave.	2.6	3.1	3.6	3.8	3.7	3.6	3.7	3.3	2.7	2.4	2.2	2.2	3.1
Station 4: Falcon Dam, Texas													
1980 Ave.	3.1	3.0	3.3	4.0	4.4	6.0	5.0	4.7	3.7	3.3	3.1	2.7	3.9
1950-1980 Ave.	3.8	4.3	4.8	5.4	5.5	5.8	6.0	5.1	3.9	3.4	3.7	3.4	4.6

Source: International Boundary and Water Commission of the United States and Mexico--United States Section, *Flow of the Rio Grande and Related Data, 1980*, Water Bulletin 50 (El Paso: IBWC, 1980), p. 148.

Table 1.4
Humidity in the Rio Grande/Río Bravo Basin
(percent humidity)

Segment	Jan	Feb	Mar	Apr	May	Jun	Jul	Aug	Sep	Oct	Nov	Dec	Annual
Station 1: Amistad Dam, Texas													
1980 Ave.	70.1	57.2	45.4	41.6	66.1	58.1	51.7	68.4	71.9	61.7	72.4	78.3	61.9
1963-1980 Ave.	61.8	59.6	53.8	58.0	64.7	62.4	59.6	61.4	67.0	66.5	65.3	63.1	61.9
Station 2: Eagle Pass, Texas													
1980 Ave.	87.3	54.1	43.7	41.7	59.2	50.4	46.9	58.8	60.1	51.6	62.1	65.4	56.8
1964-1980 Ave.	65.4	60.8	57.1	60.7	66.9	64.3	61.0	63.3	69.4	68.3	68.1	66.3	64.3
Station 3: Falcon Dam, Texas													
1980 Ave.	78.7	73.6	66.2	63.4	76.1	67.9	65.9	71.4	71.9	71.9	77.0	82.3	72.2
1950-1980 Ave.	66.6	63.7	61.7	62.1	64.9	63.7	60.6	61.7	66.6	66.4	66.5	65.9	64.2
Station 4: Nueva Ciudad Guerrero, Tamaulipas													
1980 Ave.	76.0	69.0	58.0	50.0	64.0	52.0	57.0	61.0	62.0	62.0	68.0	72.0	63.0
1961-1980 Ave.	78.0	76.0	70.0	70.0	75.0	74.0	72.0	72.0	78.0	76.0	76.0	78.0	75.0

Source: International Boundary and Water Commission of the United States and Mexico--United States Section, *Flow of the Rio Grande and Related Data, 1980*, Water Bulletin 50 (El Paso: IBWC, 1980), p. 144-45.

Table 1.5

Evaporation and Precipitation at IBWC Stations (inches)

Texas

Location	1980 Precipitation	1980 Evaporation	1980 Evaporation/ Precipitation (%)	Long-term Precipitation	Long-term Evaporation	Long-term Evaporation/ Precipitation (%)
Station 1: Presidio	11.73	91.93	7.8	8.11	98.69	12
Station 2: Johnson Ranch	8.85	83.65	10	7.73	95.67	8
Station 3: Martin King Ranch	12.75	88.95	14	14.52	79.85	18
Station 4: Long Ranch	10.41	73.30	14	17.46	59.09	29
Station 5: Amistad Dam	14.29	111.19	13	19.19	105.06	18

Table 1.5 (continued)

Location	1980 Precipitation	1980 Evaporation	1980 Evaporation/ Precipitation (%)	Long-term Precipitation	Long-term Evaporation	Long-term Evaporation/ Precipitation (%)
Station 6: Eagle Pass	14.73	98.06	15	23.00	78.16	29
Station 7: Falcon Dam (2-foot Pan)	10.89	90.00	12	20.46	88.82	23
Station 8: Falcon Dam (4-foot Pan)	10.89	111.45	10	20.46	109.31	19
Station 9: Brownsville	30.48	93.07	3	32.30	56.70	1.7

Source: International Boundary and Water Commission of the United States and Mexico--United States Section, *Flow of the Rio Grande and Related Data, 1980*, Water Bulletin 50 (El Paso: IBWC, 1980), pp. 144-45.

N/R = not reported

Table 1.6

Evaporation and Precipitation at IBWC Stations (inches)

Mexico

Location	1980 Precipitation	1980 Evaporation	1980 Evaporation/ Precipitation (%)	Long-term Precipitation	Long-term Evaporation	Long-term Evaporation/ Precipitation (%)
Station 10: Ciudad Juarez	8.43	85.91	10	9.00	92.84	10
Station 11: Ojinaga	13.59	105.16	13	9.73	103.63	9
Station 12: Ciudad Acuña	12.61	94.91	13	8.71	92.91	20
Station 13: La Amistad	9.02	N/R	N/R	14.36	106.54	13
Station 14: Jimenez	16.77	104.14	16	19.83	86.85	23

Table 1.6 (continued)

Location	1980 Precipitation	1980 Evaporation	1980 Evaporation/ Precipitation (%)	Long-term Precipitation	Long-term Evaporation	Long-term Evaporation/ Precipitation (%)
Station 15: Hidalgo	20.08	100.71	20	19.51	105.46	18
Station 16: Nuevo Laredo	17.51	140.70	12	19.74	108.16	18
Station 17: Nuevo Cd. Guerrero	11.05	94.46	12	19.65	94.32	21
Station 18: Ciudad Mier	13.14	106.23	12	23.66	101.96	23
Station 19: Retamal	17.65	77.69	22	21.69	81.81	26

Source: International Boundary and Water Commission of the United States and Mexico--United States Section, *Flow of the Rio Grande and Related Data, 1980,* Water Bulletin 50 (El Paso: IBWC, 1980), pp. 115-33, 144-45.

N/R = not reported

Table 2.1

Rio Grande/Rio Bravo Basin Segments: TDWR Classification and Ranking

Segment Number	Description	Location	Length (miles)	Class	Rank
2301	Rio Grande Tidal	Gulf of Mexico 2.9 miles south of SH4 in Cameron County to a point 6.7 miles downstream from the international Bridge in Brownsville	49.3	EL	106
2302	Rio Grande River	from a point 6.7 miles downstream from the international Bridge in Brownsville to Falcon Dam	230.9	EL	108
2303	Falcon Reservoir	from Falcon Dam to the confluence of the Arroyo Salado from Mexico south of San Ygnacio in Zapata County	68.3	EL	272
2304	Rio Grande River	Falcon Lake headwater of the confluence of the Arroryo Salado from Mexico south of San Ygnacio in Zapata County to Amistad Dam	226.5	EL	83
2305	Amistad Reservoir	from Amistad Dam to a point 3.7 miles south of U.S. 90 and 8.8 miles east of the Val Verde-Terrell County line in Val Verde County	74.7	EL	291

Table 2.1 (continued)

Segment Number	Description	Location	Length (miles)	Class	Rank
2306	Rio Grande River	Amistad Reservoir headwater at a point 3.7 miles south of U.S. 90 and 8.8 miles east of the Val Verde-Terrell County line in Val Verde County to the Rio Conchos (Mexico) confluence near Presidio in Presidio County	312.8	EL	227
2307	Rio Grande River	Rio Conchos (Mexico) confluence near Presidio to Riverside Diversion Dam	222.0	EL	720
2308	Rio Grande River	Riverside Diversion Dam near Yaleta to New Mexico	35.8	WQL	29
2309	Devils River	Amistad Reservoir headwater to River headwater at a point 4.4 miles south of FM Road 1828, 2.8 miles north of SH 29, and .9 miles east of 100 degrees 45 minutes longitude in Schleicher County	136.6	EL	269

Table 2.1 (continued)

Segment Number	Description	Location	Length (miles)	Class	Rank
2310	Pecos River	Amistad Reservoir headwaters at the 1117' contour line 1.5 miles north of U.S. 90 in Val Verde County to the county road low water crossing near Pandale	49.0	EL	279
2311	Pecos River	county road low water crossing near Pandale to Red Bluff Dam	362.9	EL	81
2312	Red Bluff Reservoir	from Red Bluff Dam to a point on the Texas-New Mexico State line 5 miles north of U.S. 285 on the Loving-Reeves county line	11.4	EL	243
2313	Arroyo Colorado	from entrance of Laguna Madre between Willacy and Cameron Counties to Highway 1016, 2 miles south of Mission in Hidalgo County	80.9	WQL	18

Source: Texas Department of Water Resources, *The State of Texas Water Quality Inventory*, 6th ed., Report LP-59 (Austin: TDWR, 1982).

Table 2.2

TDWR Water Quality Standards and Uses for Rio Grande/Rio Bravo Basin Segments
(maximum values in milligrams per liter unless indicated)

Segment	2301	2302	2303	2304	2305	2306	2307
Designated uses*	2,3	1,2,3,4	1,2,3,4	1,2,3,4	1,2,3,4	1,2,3,4	1,2,3,4
Chloride	-	270	200	200	150	200	300
Sulfate	-	350	250	300	250	500	550
Total dissolved solids	-	880	700	1,000	500	1,200	1,500
Dissolved oxygen	5.0	5.0	5.0	5.0	5.0	5.0	5.0
pH range	6.5-9.0	6.5-9.0	6.5-9.0	6.5-9.0	6.5-9.0	6.5-9.0	6.5-9.0
Fecal coliform	1,000	200	200	200	200	200	200
Temperature F	95	90	93	95	88	93	93

Table 2.2 (continued)

Segment	2308	2309	2310	2311	2312	2201
Designated Uses*	2,3,4	1,2,3,4	1,2,3,4	1,2,3	1,2,3	2,3
Chloride	500	20	1,000	7,000	6,000	-
Sulfate	700	20	500	3,500	3,500	-
Total dissolved solids	1,800	300	3,000	15,000	15,000	-
Dissolved oxygen	5.0	6.0	5.0	5.0	5.0	4.0
pH range	6.5-9.0	6.5-9.0	6.5-9.0	6.5-9.0	6.5-9.0	6.5-9.0
Fecal coliform	2,000	200	200	200	200	2,000
Temperature F	95	90	92	92	90	95

Source: Texas Department of Water Resources, *Texas Surface Water Quality Standards*, LP-71 (Austin: TDWR, April 1981).

* indicates minimum level requirement for each of the following uses:

1. Contact Recreation
2. Non-Contact Recreation
3. Propagation of Fish and Wildlife
4. Domestic Raw Water Supply

Table 2.3

**Texas Department of Health Chemical Quality Standards
(maximum values in milligrams per liter)**

Contaminants	Community System	Noncommunity
Alkalinity	**	
Arsenic	0.05	
Barium	1.0	
Cadmium	0.01	
Chlorides*	300	300
Chlorine residual	**	
Chromium	0.05	
Copper	1.0	1.0
Corrosivity	noncorrosive	noncorrosive
Total dissolved solids*	1000	1000
Fluoride*	1.4-1.8	
Foaming agents	0.5	0.5
Lead	0.05	0.05

Table 2.3 (continued)

Contaminants	Community System	Noncommunity System
Iron	0.3	0.3
Manganese	0.05	0.05
Mercury	0.002	
Nitrate	10 (N)	10 (N)
pH (pH units)	> 7.0	> 7.0
Selenium	0.01	
Silver	0.05	
Sulfate	300	300
Zinc	5.0	5.0

Source: Texas Department of Health, Division of Water Hygiene, *Drinking Water Standards Governing Drinking Water Quality and Reporting Requirements for Public Water Supply Systems* (Austin: Texas Department of Water Resources, November 29, 1980).

* indicates monitoring requirement only.

**indicates a standard which differs from EPA regulations.

Table 2.4

SARH Water Use Classifications

Class	Definition
DA	Domestic and industrial water supply using disinfection treatment Contact water recreation
DI	Domestic and industrial water supply using conventional treatment
DII	Water for recreation, conservation of flora and fauna, or industrial use
DIII	Water for agricultural or industrial use
DIV	Water for industrial use (excluding food processing)

Source: Secretaría de Agricultura y Recursos Hidráulicos, *Reglamento para la Prevención y Control de la Contaminación de Aguas* (Mexico, D.F.: SARH, March 29, 1973).

Table 2.5

SARH Water Quality Standards: General Contaminants
(maximum values in milligrams per liter unless indicated)

Contaminant	DA	DI
pH (pH units)	6.5-8.5	6.0-9.0
Temperature (Celsius) #	30	30
Dissolved oxygen *	4.0	4.0
coliform (colonies per 100 ml)	200 in no more than 10% of monthly samples) 2,000 (maximum)	1,000 (in no more than 10% of monthly samples) 2,000 (maximum)
Oils and greases	76	10
Dissolved solids	1,000	1,000
Turbidity (Jackson units)	10	natural condition
Color (platinum cobalt scale)	20	must coagulate with conventional treatment
Odor and taste	absent	must be removable with conventional treatment
Nitrogen and phosphorus	should not hyperfertilize	should not hyperfertilize
Floating material	absent	absent

Table 2.5 (continued)

DII	DIII	DIV
6.0-9.0	6.0-9.0	5.0-9.5
30	30	-
4.0	3.2	3.2
10,000 (maximum monthly sample) 20,000 (maximum)	1,000 +	-
visibly absent	-	-
2,000	2,000	-
natural condition	natural condition	-

Table 2.5 (continued)

DII	DIII	DIV
natural condition	natural condition	-
natural condition	-	-
should not hyperfertilize	should not hyperfertilize	should not hyperfertilize
absent	absent	absent

Source: Secretaría de Agricultura y Recursos Hidráulicos, Reglamento para la Prevención y Control de la Contaminación de Aguas (Mexico, D.F.: SARH, March 29, 1973).

* Minimum level.
# Temperature maximum not applicable if violations are due to natural causes.
+ Limitation applicable only to irrigation of (a) produce to be consumed without boiling or (b) fruits with direct soil contact.

Table 2.6

SARH Water Quality Standards: Toxic Contaminants
(maximum values in milligrams per liter unless indicated)

Contaminant	DA	DI	DII	DIII	DIV
Arsenic	0.05	0.05	1.0	5.0	-
Barium	1.0	1.0	5.0	-	-
Boron	1.0	1.0	-	2.0	-
Cadmium	0.01	0.01	0.01	0.005	-
Copper	1.0	1.0	0.1	1.0	-
Mercury	0.005	0.005	0.01	-	-
Lead	0.05	0.05	0.10	-	-
Selenium	0.01	0.01	0.05	0.05	-
Cyanide	0.20	0.20	0.02	-	-
Phenol	0.001	0.001	1.0	-	-

Table 2.6 (continued)

Contaminant	DA	DI	DII	DIII	DIV
Methylene blue Active Substances	0.50	0.50	3.0	-	-
Chloroform	0.15	0.15	-	-	-
Aldrin	0.017	0.017	-	-	-
Chlorodane	0.003	0.003	-	-	-
DDT	0.042	0.042	-	-	-
Dieldrin	0.017	0.017	-	-	-
Heptachlor	0.018	0.018	-	-	-
Lindane	0.056	0.056	-	-	-
Methoxyclor	0.035	0.035	-	-	-
Organic phosphates	0.10	0.10	-	-	-

Table 2.6 (continued)

Contaminant	DA	DI	DII	DIII	DIV
Toxaphene	0.005	0.005	-	-	-
Herbicides (total)	0.10	0.10	-	-	-
Beta radiation*	1.0	1.0	1.0	-	-
Radium 226*	3.0	3.0	3.0	-	-
Strontium	10.0	10.0	10.0	-	-

Source: Secretaría de Agricultura y Recursos Hidráulicos, *Reglamento para la Prevención y Control de la Contaminación de Aguas* (Mexico, D.F.: SARH, March 29, 1973).

* in pica curies per liter.

Table 2.7

SARH Water Quality Regulations: Emission Restrictions

Contaminant	Limitation
Settleable solids	1.0 mg/l
Greases and oils	70 mg/l
Floating material	no materials that can be retained by a 3 mm net
Temperature	35 degrees Celsius
pH	4.5 to 10.0 pH units

Source: Secretaría de Agricultura y Recursos Hidráulicos, *Reglamento para la Prevención y Control de la Contaminación de Aguas* (Mexico, D.F.: SARH, March 29, 1973).

Table 2.8

SSA Drinking Water Regulations
(maximum values in milligrams per liter unless indicated)

Class	Contaminant	Limitation
Physical	Turbidity	10
	Color (platinum cobalt scale)	20
	pH	6.0 - 8.0
	Odor	absence
	Taste	agreeable
Chemical	Nitrogen Ammonia	0.5
	Proteiform nitrogen (N)	0.10
	Nitrogen (nitrites)	0.05
	Nitrogen (nitrates)	5.0
	Oxygen minimum	3.0
	Total solids (TS)	Preference: TS ≤ 500
		Maximum: TS ≤ 1,000
	Total alkalinity (CaCO3)	400.0
	Total hardness (CaCO3)	300.0
	Hardness (CaCO3) in natural waters	150.0
	Chloride (Cl)	250.0
	Sulfates (SO4)	250.0
	Magnesium (Mg)	125.0
	Zinc (Zn)	15.0
	Copper (Cu)	3.0
	Fluoride (F)	1.5
	Iron (Fe) and Manganese (Mn)	0.3
	Lead (Pb)	0.1
	Arsenci (As)	0.05
	Selinium (Se)	0.05
	Chromium (Cr)	0.05
	Compound Fenol	0.001
	Free Chlorine (FC)	$0.2 \leq FC \leq 1.0$
		greater than 1.0

Source: Felipe Ochoa y Asociados, S.C., *Estudio de la Calidad del Agua en la Cuenca del Río Bravo* (Mexico, D.F.: Secretaría de Agricultura y Recursos Hidráulicos, September 1978). (Limited Distribution Document.)

Table 2.9

**Water Quality Standards in the United States and Mexico
Decentralized versus Centralized Administrations**

Topic	United States	Mexico
Surface water quality	**FEDERAL**	**FEDERAL**
	Laws: Federal Water Pollution Control Act (P.L. 92-500)	**Laws:** Ley Federal para Prevenir y Controlar la Contaminación Ambiental (3/23/71)
	Clean Water Act (P.L. 95-217)	Ley Federal de Protección al Ambiente (12/22/81)
		Ley Federal de Aguas
	Regulations:	**Regulations:**
	Various EPA regulations	SARH's Reglamento para la Prevención y Control de la Contaminación de Aguas (3/29/73)
	Enforcement: EPA	**Enforcement: SARH**
	TEXAS	**STATES**
	TDWR regulates and enforces the standards	Cooperate in regulation and enforcement

Table 2.9 (continued)

Groundwater quality	**FEDERAL** **Laws:** Federal Water Pollution Control Act (P.L. 92-500) authorized emergency action No standards set	**FEDERAL** **Laws:** Ley Federal para Prevenir y Controlar la Contaminación Ambiental authorizes emergency actions No standards set
Drinking water quality	**FEDERAL** **Laws:** Safe Drinking Water Act (P.L. 93-523) **Regulations: EPA** **Enforcement: Delegated to the states** **TEXAS** Responsible for system regulation and enforcement **LOCAL INSTITUTIONS** Local institutions manage water systems	**FEDERAL** **Laws:** Sanitary Code of the United States of Mexico **Regulations: SSA** **Enforcement: SSA** **STATES** Cooperate in regulation and enforcement **LOCAL GOVERNMENTS** May operate water systems May enforce drinking water regulations

Table 2.10

Rio Grande/Río Bravo Basin Surface Water Quality Problem Areas
and Ranking

Location	Segment number	Contaminants	Possible Causes
El Paso	2308	TDS, fecal coliforms, sulfates, chlorides	irrigation return flows municipal discharges, lack of flow in fall and winter
Fort Quitman	2307	TDS	irrigation return flows
Above Rio Conchos	2307	TDS	-
Rio Conchos	N/A	fecal coliform, sulfate	-
Below Rio Conchos	2306	fecal coliform, DDT, endrin	-
Del Rio/Ciudad Acuña	2304	fecal coliform	municipal discharges
Laredo/Nuevo Laredo	2304	fecal coliform	municipal discharges
Red Bluff Reservoir	2312	TDS, oil	natural conditions, oil spills
Pecos River	2311	TDS, dissolved oxygen, oil	salt deposits, irrigation return flows, inflow of brine from oilfields

Table 2.10 (continued)

Pecos at Langtry	2310	TDS	natural conditions
Nuevo Guerrero, Mexico	2302	suspended silt, phosphates, sodium, grease, oil	municipal discharges
Morillo Drain	2302	TDS	irrigation return flows
Rio San Juan	N/A	TDS, phosphorus, grease, oil	industrial discharges
Reynosa, Andalduas Dam, Rio Bravo	2302	suspended silt, chlorides, oil	municipal and industrial discharges
Brownsville/Matamoros	2301	dissolved oxygen fecal coliform, pesticides	municipal discharges, pesticide use
Arroyo Colorado	2201	dissolved oxygen, fecal coliform, pesticides, some toxic chemicals	municipal discharges, irrigation return flows

Source: Black and Veach, *Toxic Substances Sampling Program* (Dallas: Lower Rio Grande Development Council, June 1981).

Felipe Ochoa y Asociados, *Estudio de la Calidad del Agua en la Cuenca del Río Bravo* (Mexico, D.F.: Secretaría de Agricultura y Recursos Hidráulicos, September 1978). (Limited Distribution Document.)

U. S. Geological Survey, *Water Resources Data for Texas: Water Year 1980*, vol. 3 (Austin: U.S. Geological Survey, 1981).

U. S. Section, International Boundary and Water Commission, *Flow of the Rio Grande and Related Data: 1980*, Water Bulletin 50 (El Paso: IBWC, 1981).

Table 3.1

Major and Minor Aquifers of the Rio Grande/Río Bravo Basin

Major Aquifer	Minor Aquifer
Mesilla Bolson	Salt Bolson
Hueco Bolson	Capitan Limestone
Red Light Draw Bolson	Bone Spring Limestone
Green River Valley Bolson	Victoria Peak Limestone
Presidio Bolson	Santa Rosa
Redford Bolson	Marathon Limestone
Edwards-Trinity (Plateau)	Rustler
Carrizo-Wilcox	Igneous Rock
Gulf Coast	Cenozoic Alluvium

Source: Randall J. Charbeneau, "Groundwater Resources of the Texas Rio Grande Basin," *Natural Resources Journal 22*, no. 4 (October 1982): 957.

Table 3.2

Storage Capacity and Recharge Volume of
Rio Grande/Río Bravo Basin Aquifers
(millions of cubic meters)

Aquifer	Aquifer Storage	Annual Recharge
Mesilla Bolson	690.8	22.2
Hueco Bolson	13,075.0	7.4
Red Light Draw Bolson	555.1	2.5
Green River Valley Bolson	259.0	1.2
Presidio and Redford Bolsons	925.1	8.6
Edwards-Trinity (Plateau)	131,009.0	633.9
Carrizo-Wilcox	197.4	16.9
Gulf Coast	N/A	14.1

Source: Randall J. Charbeneau, "Groundwater Resources of the Texas Rio Grande Basin," *Natural Resources Journal 22*, no. 4 (October 1982): 960.

N/A = not available

Table 3.3

Annual Volume of Groundwater Available in the Rio Grande/Río Bravo Basin Aquifers
(millions of cubic meters)

Aquifer	1980-89	1990-99	2000-09	2010-19	2020-29	2030
Mesilla Bolson	40.6	51.1	51.1	51.1	48.7	46.6
Hueco Bolson	111.8	153.8	206.4	289.6	375.4	460.7
Red Light Draw Bolson	12.7	12.7	12.7	12.7	12.7	2.5
Green River Valley Bolson	6.9	6.0	6.0	6.0	6.0	1.2
Presidio and Redford Bolsons	25.8	25.8	25.8	25.8	25.8	8.6
Edwards-Trinity (Plateau)	633.9	633.9	633.9	633.9	633.9	633.9
Carrizo-Wilcox	20.2	20.2	20.2	20.2	20.2	16.9
Gulf Coast	14.1	14.1	14.1	14.1	14.1	14.1
Total	865.1	917.6	970.2	1053.4	1136.8	1184.5

Source: Texas Department of Water Resources, *Groundwater Availability in Texas*, Report 238 (Austin: TDWR, September 1979), p. 71.

* Pumpage from the aquifer directly depletes surface supplies.

Table 3.4

Groundwater Volume Withdrawn from the Lower Mesilla Bolson
(millions of cubic meters)

Use	Annual Water Withdrawn	
	1957-1975	1975
Municipal use:		
Upper aquifer	6.8	6.2
Medium aquifer	4.3	1.7
Deep aquifer	11.3	15.7
Agricultural use:		
All-wet period	3.7-6.2*	37.0
All-drought period	62.0*	99.0
Other uses:	N/A	9.9

Source: Randall J. Charbeneau, "Groundwater Resources of the Texas Rio Grande Basin," *Natural Resources Journal 22*, no. 4 (October 1982): 959.

* These data were taken from the period 1973 to 1975.

N/A = not available.

Table 3.5

Stored Groundwater Volume of the West
Rio Grande/Río Bravo Basin
(millions of cubic meters)

Location	Water Stored
East Franklin Mountains	**13,075**
El Paso Valley	**495-990**
East of Franklin and Oregon Mountains, New Mexico	**7,650**
Ciudad Juarez, Chihuahua	**4,930**

Source: Texas Department of Water Resources, *Availability of Fresh and Slightly Saline Ground Water in the Basin of Westernmost Texas*, Report 256 (Austin: TDWR, September 1980), p. 93.

Table 3.6

Volume of Present and Projected Water Use
in the El Paso Area
(millions of cubic meters)

Year	Total Use	Surface Water	Ground-water	Mesilla Bolson	Hueco Bolson
1974	138.0	17.5	120.5	29.7	90.8
1980	169.0	16.7	152.3	40.6	111.8
1990	222.0	17.1	204.9	51.1	153.8
2000	278.3	20.8	257.4	51.1	206.4
2010	362.8	22.1	340.7	51.1	286.6
2020	447.1	23.1	424.1	48.7	375.4
2030	532.1	24.8	507.3	46.6	560.7

Source: Randall J. Charbeneau, "Groundwater Resources of the Texas Rio Grande Basin," *Natural Resources Journal* 22, no. 4 (October 1982): 957.

Table 3.7

Storage and Recharge Volumes of the Green River Valley and Presidio and Redford Bolsons (millions of cubic meters)

Aquifer	Recoverable Storage	Annual Recharge
Green River Valley	259	1.2
Presidio and Redford	925	8.6

Source: Texas Department of Water Resources, *Groundwater Availability in Texas*, Report 238 (Austin: Texas Department of Water Resources, September 1979), p. 34.

Table 3.8

Water Volume Statistics for Aquifers Located on the Mexican Side of the Rio Grande/Río Bravo Basin (millions of cubic meters)

Location	Water Withdrawn	Annual Recharge	Storage Volume
Aldama (Chihuahua)	54	65	2,560
Cuauhtemoc (Chihuahua)	N/A	N/A	2,400
Delicias (Chihuahua)	125	250	5,600
Jimenez-Camargo (Chihuahua)	370	N/A	3,400
Juarez Valley (Chihuahua)	144	167	N/A
Acuña (Coahuila)	400-500	400-500	N/A
Saltillo Ramos-Arizpe (Coahuila)	27	30-50	N/A
Portion of Nuevo León	144	865	*
Bajo Río Bravo and Bajo Río San Juan (Tamaulipas)	40	35	N/A
Total	1,604-1,704	1,812-1,972	N/A

Source: Secretaría de Agricultura y Recursos Hidráulicos, *Esquema de Desarrollo Hidráulico para la Cuenca del Río Bravo* (México, D.F.: Secretaría de Agricultura y Recursos Hidráulicos, 1980), p. 46. (Internal Document.)

* Estimated as 1,050 million cubic meters for each meter of depression.

N/A = not available

Table 3.9

**Constituents in the Rio Grande Alluvium and the Hueco
and Mesilla Bolsons
(concentrations in mg/1 except as noted otherwise)**

Constituent	Rio Grande Alluvium	Hueco Bolson	Mesilla Bolson
Silica	0.25-70	0-57	6.3-50
Iron	0-7.2	0-188	0-3.1
Calcium	4-1,492	4-7,612	2.4-152
Magnesium	0-404	0-180	0-80
Sodium	100-3,183	63-85,148	2.5-1,790
Potassium	4-679	1.6-1,982	0.5-28
Bicarbonate	0-618	17-660	52-597
Sulfate	9-4,240	1-5,800	67-2,470
Chloride	32-11,400	30-150,560	29-2,120
Fluoride	0.03-3	0-5	0.3-5.6
Nitrate	< 0.05-73	0-35	0-13
Boron	0.11-0.29	0.05-0.42	0.08-0.55
Dissolved solids	263-23,680	117-246,510	252-4,542
Hardness	8-14,417	0-28,000	6-708
Specific conductance *	458-11,000	8-60,800	408-7,390
pH *	6.1-9.3	6.4-9.7	7.0-9.3

Source: Texas Department of Water Resources, *Groundwater Development in the El Paso Region, Texas with Emphasis on the Lower El Paso Valley*, Report 246 (Austin: Texas Department of Water Resources, June 1980), p. 39.

* pH is expressed in pH units. Specific conductance is measured in micromhos per centimeter.

Table 3.10

Dissolved Solids in Pecos County Groundwater
(concentration expressed in mg/l)

Irrigation Areas	Concentration
West South Coyanosa	less than 500
East of South Coyanosa	greater than 1,000
Fort Stockton-Leon-Belding	2,000 - 4,000
Girvin	3,250 - 5,000
Bakersfield	2,000

Source: Texas Department of Water Resources, *Occurrence and Quality of Groundwater in the Edwards-Trinity (Plateau) Aquifer in the Trans-Pecos Region of Texas,* Report 255 (Austin: TDWR, September 1980), p. 15.

Table 4.1

Population of Sixteen Texas Border-Region Counties, 1930-80

County	1930	1940	1950	1960	1970	1980	Rate*
Brewster	6,624	6,478	7,309	6,434	7,780	7,573	0.26
Cameron	77,540	83,202	125,170	151,098	140,368	209,727	1.99
Culberson	1,228	1,653	1,825	2,794	3,429	3,315	1.99
Dimmit	8,828	8,542	10,654	10,095	9,039	11,367	0.51
El Paso	131,597	131,067	194,968	314,070	359,291	479,899	2.59
Hidalgo	77,004	106,059	106,446	180,904	181,535	283,229	2.60
Hudspeth	3,728	3,149	4,298	3,343	2,392	2,728	-0.62
Jeff Davis	1,800	2,375	2,090	1,582	1,527	1,647	-0.18
Kinney	3.980	4,533	2,668	2,452	2,006	2,279	-1.12
Maverick	6,120	10,071	12,292	14,508	18,093	31,398	3.27
Presidio	10,154	10,925	7,354	5,460	4,842	5,188	-1.34
Starr	11,409	13,312	13,948	17,137	17,707	27,266	1.74
Terrell	2,666	2,952	3,189	2,600	1,940	1,595	-1.03
Val Verde	14,924	15,453	16,635	24,461	27,471	35,910	1.76
Webb	42,128	45,916	56,141	64,791	72,859	99,258	1.71
Zapata	2,867	3,916	4,405	4,393	4,352	6,628	1.68
Total	402,597	449,603	569,342	806,122	854,631	1,209,007	2.20

Source: U.S. Bureau of the Census, *1980 Census of Population.* Vol. 1, Characteristics of the Population, Chapter A, Number of Inhabitants, Part 45, Texas (Washington, D.C.: U.S. Government Printing Office, March 1982).

* Continuous annual rate of increase, 1930-80.

Table 4.2

Population Growth 1930-1980
The Rio Grande/Río Bravo Basin, Texas, and the United States

Region	1930	1940	1950	1960	1970	1980	Rate (1930-1980)	Rate (1960-1980)
Rio Grande/ Río Bravo Basin	410,079	479,290	576,188	814,425	861,649	1,209,007	2.16	1.98
Texas	5,825,715	6,414,824	7,711,194	9,579,677	11,196,730	14,229,191	1.79	1.98
U.S.	122,775,046	131,669,275	151,325,798	179,323,175	203,302,031	226,504,825	1.22	1.17

Source: U.S. Bureau of the Census, *1980 Census of Population*, Vol. 1, Characteristics of the Population, Chapter A, Number of Inhabitants, Part 45, Texas (Washington, D.C.: U.S. Government Printing Office, March 1982).

Texas 2000 Commission, *Texas Trends*, (Austin: Texas 2000 Commission, Office of the Governor, 1980).

U.S. Bureau of the Census, *Statistical Abstract of the United States, 1981*, (Washington, D.C.: U.S. Government Printing Office, 1981).

Table 4.3

Border-Region County Areas, Populations, and Densities

County	Area (sq mi)	Population	Persons/sq mi
Brewster	6,169	7,573	1.2
Cameron	905	209,727	231.7
Culberson	3,815	3,315	0.9
Dimmit	1,307	11,367	8.7
El Paso	1,014	479,899	473.3
Hidalgo	1,569	283,229	180.5
Hudspeth	4,566	2,728	0.6
Jeff Davis	2,258	1,647	0.7
Kinney	1,359	2,279	1.7
Maverick	1,287	31,398	24.4
Presidio	3,857	5,188	1.3
Starr	1,226	27,266	22.2
Terrell	2,357	1,595	0.7
Val Verde	3,150	35,910	11.4
Webb	3,363	99,258	29.5
Zapata	999	6,628	6.6
Total	39,500	1,209,007	30.6
Texas	267,339	14,229,191	53.2
U.S.	3,618,770	226,504,825	62.6

Sources: U.S. Bureau of the Census, *1980 Census of Population*, Vol.1, Characteristics of the Population, Chapter A Number of Inhabitants, Part 45, Texas, (Washington, D.C.: U.S. Government Printing Office, March 1982).

U.S. Bureau of the Census, *Statistical Abstract of the United States 1981*, (Washington, D.C.: U.S. Government Printing Office, 1981).

These population densities are calculated using total surface areas (land area plus water surface area). The actual number of inhabitants per land area is therefore slightly larger.

Table 4.4

Selected Population Characteristics of Texas

County	Total	Population Male	Female	Median Age	Sex Ratio	Persons of Spanish Origin	Percent of Spanish Origin
Brewster	7,753	3,769	3,804	27.1	.991	3,262	43.1
Cameron	209,727	100,581	109,146	25.0	.922	161,654	77.1
Culberson	3,345	1,599	1,716	24.3	.932	2,101	63.4
Dimmit	11,367	5,588	5,779	24.7	.967	8,845	77.8
El Paso	479,899	235,043	244,856	25.0	.960	297,001	61.9
Hidalgo	283,229	135,917	147,312	24.1	.923	230,212	81.3
Hudspeth	2,728	1,416	1,312	25.5	1.079	1,589	58.2
Jeff Davis	1,647	814	833	33.0	.977	777	47.2

Table 4.4 (continued)

County	Total	Population Male	Female	Median Age	Sex Ratio	Persons of Spanish Origin	Percent of Spanish Origin
Kinney	2,279	1,139	1,140	32.4	.999	1,310	57.5
Maverick	31,398	15,048	16,350	22.3	.920	28,366	90.3
Presidio	5,188	2,498	2,690	29.7	.929	3,989	76.9
Starr	27,266	13,258	14,008	22.8	.946	26,428	96.9
Terrell	1,595	827	768	31.8	1.077	691	43.3
Val Verde	35,910	17,794	18,116	24.6	.982	22,601	62.9
Webb	99,258	47,435	51,823	23.6	.915	90,842	91.5
Zapata	6,628	3,310	3,318	29.6	.998	5,042	76.1
Total	1,209,007	586,036	622,971	N/A	.941	884,710	73.2

Sources: U.S. Bureau of the Census, *1980 Census of Population*, State Tape file 1 compiled by the Bureau of Business Research, University of Texas at Austin Data Center, (Washington, D.C.: U.S. Government Printing Office, 1982).

N/A = not available.

Table 4.5

Population and Housing Data of Texas Border Counties

County	Population	Housing Units	Households	Household Mean	Size Median
Brewster	7,573	3,247	2,694	2.63	2.21
Cameron	209,727	65,970	58,418	3.56	3.14
Culberson	3,315	1,178	987	3.35	3.03
Dimmit	11,367	3,646	3,135	3.58	3.25
El Paso	479,899	147,964	140,806	3.32	3.02
Hidalgo	283,229	89,047	75,816	3.71	3.33
Hudspeth	2,728	1,145	822	3.30	3.02
Jeff Davis	1,647	963	592	2.75	2.32
Kinney	2,279	1,101	771	2.96	2.32
Maverick	31,398	8,669	7,583	4.05	3.78
Presidio	5,188	2,115	1,680	3.08	2.52
Starr	27,266	7,833	6,858	3.94	3.66
Terrell	1,595	949	570	2.80	2.41
Val Verde	35,910	12,261	10,355	3.33	3.03
Webb	99,258	27,753	25,896	3.79	3.48
Zapata	6,628	3,054	2,059	3.22	2.56
Total	1,209,007	376,895	339,042	3.57	N/A

Source: U.S. Bureau of the Census, *1980 Census of Population*, State Tape File 1 compiled by the Bureau of Business Research, University of Texas at Austin Data Center, (Washington, D.C.: U.S. Government Printing Office, 1982).

N/A = not available

Table 4.6

Texas Vital Statistics, 1981

County	Birth Rate	Death Rate	Natural Increase	Projected 1980-81 Population Change
Brewster	1.65	0.74	0.91	-0.16
Cameron	2.57	0.59	1.98	4.47
Culberson	2.69	0.77	1.92	4.10
Dimmit	2.07	0.50	1.56	2.54
El Paso	2.25	0.51	1.74	3.33
Hidalgo	2.64	0.55	2.09	4.90
Hudspeth	1.80	0.50	1.30	3.70
Jeff Davis	1.51	0.72	0.78	1.64
Kinney	1.43	1.08	0.35	2.50
Maverick	2.47	0.45	2.02	6.47
Presidio	1.68	0.82	0.86	1.37
Starr	2.59	0.60	1.99	5.57
Terrell	1.31	1.25	0.06	0.88
Val Verde	2.52	0.63	1.88	3.32
Webb	2.72	0.61	2.11	4.59
Zapata	2.19	0.76	1.44	3.62
Texas	1.96	0.77	1.19	2.42
U.S.	1.62	0.89	0.73	N/A

Source: Texas Department of Health, "Population Data System," Austin, 1982. (Unpublished Computer Printout.)

Texas Department of Health, *Texas Vital Statistics 1981*, (Austin: TDH, 1982).

U.S. Bureau of the Census, *USA Statistics in Brief*, (Washington, D.C.: U.S. Government Printing Office, 1982).

Totals may not equal sums due to rounding.

N/A = not available.

Table 4.7

Population and Growth Rates 1960-80
Four Mexican States and Mexico

State	1960	1970	1980	%Increase (1960-80)	Rate
Chihuahua	1,227,000	1,613,000	1,933,856	57.6	2.28
Coahuila	908,000	1,115,000	1,588,401	71.6	2.80
Nuevo León	1,079,000	1,695,000	2,463,298	128.3	4.13
Tamaulipas	1,024,000	1,457,000	1,924,934	88.0	3.16
Total	4,238,000	5,880,000	7,880,489	85.9	3.10
Mexico	36,106,000	49,417,000	67,382,581	86.6	3.10

Sources: Patricia M. Rowe, *Country Demographic Profiles Mexico*, publication of the U.S. Bureau of the Census (Washington, D.C.: U.S. Government Printing Office, September 1979).

Secretaría de Programmación y Presupuesto, *X Censo General de Población y Vivienda, 1980*, Resultados Preliminares a Nivel Nacional y por Entidad Federativa (Mexico, D.F.: SPP, 1981).

Table 4.8

Select Population Characteristics of Mexico

State	Total	Male	Female	Sex Ratio
Chihuahua	1,933,856	952,936	980,920	.971
Coahuila	1,558,401	782,539	775,862	1.009
Nuevo León	2,463,298	1,226,870	1,236,428	.992
Tamaulipas	1,924,934	953,773	971,161	.982
Total	7,880,489	3,916,118	3,964,371	.988
Mexico	67,382,581	33,295,260	34,087,321	.977

Sources: Secretaría de Programmación, *X Censo General de Población y Vivienda, 1980*, Resultados Preliminares a Nivel nacional y por Entidad Federativa (Mexico, D.F.: SPP, 1981).

Table 4.9

Population and Land Area: Mexico

State	Population	Land Area (sq.mi.)	Persons/sq.mi.
Chihuahua	1,933,856	94,154	20.5
Coahuila	1,558,401	57,653	27.0
Nuevo León	2,463,298	24,957	98.7
Tamaulipas	1,924,934	30,515	63.1
Total	7,880,489	207,279	38.0
Mexico	67,383,000	750,782	89.5

Sources: Secretaría de Programmación y Presupuesto, *Datos Basicos Sobre la Población de México 1980-2000* (Mexico, D.F.: SPP, 1981).

Table 4.10

Mexican Border State and City Populations

State	City	Population
Chihuahua	Ciudad Juarez	578,471
	Ojinaga*	30,644
Coahuila	Ciudad Acuña*	46,299
	Piedras Negras*	66,525
Tamaulipas	Nuevo Laredo	238,740
	Reynosa	244,913
	Matamoros	242,973
Total		1,448,565

Source: Edmundo Vicotria Mascorro, "Características Del Desarrollo Económico de la Franja Fronteriza Norte de México," *Natural Resources Journal*22, no. 4 (October 1982): 827.

* Estimates of the 1980 populaiton were derived by extrapolating 1970-80 growth rates for the states of Chihuahua and Coahuila (1.81 and 3.54 percent, respectively).

Table 4.11

TDWR Population Projections: Texas Border Region

County	1980	1990	2000	Rate
Brewster	7,753	7,421	7,668	-0.06
Cameron	209,680	305,523	399,480	3.22
Culberson	3,315	3,301	3,269	-0.07
Dimmit	11,367	14,271	17,412	2.13
El Paso	479,899	632,398	790,964	2.50
Hidalgo	283,229	431,842	599,637	3.75
Hudspeth	2,728	3,218	3,638	1.44
Jeff Davis	1,647	1,861	1,880	0.66
Kinney	2,279	2,717	3,131	1.59
Maverick	31,398	51,278	75,007	4.35
Presidio	5,188	5,855	6,444	1.09
Starr	27,266	41,406	58,269	3.80
Terrell	1,595	N/A	1,166	-1.57
Val Verde	35,910	51,528	71,440	3.44
Webb	99,258	137,123	176,069	2.87
Zapata	6,628	8,734	10,870	2.47
Total	1,209,007	1,698,476	2,226,357	3.05
Texas	14,228,383	17,817,323	21,227,222	2.00

Source: Texas Department of Water Resources, "Population Data System," Austin, November 1981. (Unpublished Computer Printout.)

Texas Department of Water Resources, "Population Data System," Austin, October 1981. (Unpublished Computer Printout.)

N/A = not available

Table 4.12

TDH Population Projections: Texas Border Region

County	1980	1990	2000	Rate
Brewster	7,573	7,754	8,151	0.37
Cameron	209,727	339,769	569,895	5.00
Culberson	3,315	6,287	15,131	7.59
Dimmit	11,367	15,158	20,084	2.85
El Paso	479,899	715,641	1,157,140	4.40
Hidalgo	283,229	476,236	826,069	5.35
Hudspeth	2,728	4,837	11,244	7.08
Jeff Davis	1,647	2,445	4,887	5.44
Kinney	2,279	3,794	8,185	6.39
Maverick	31,398	64,546	124,920	6.90
Presidio	5,188	6,256	7,836	2.06
Starr	27,266	47,913	86,867	5.80
Terrell	1,595	2,194	4,265	4.92
Val Verde	35,910	52,947	83,389	4.21
Webb	99,258	162,667	277,801	5.15
Zapata	6,628	10,091	15,127	4.13
Total	1,209,504	1,915,535	3,220,991	4.90
Texas	14,229,191	19,197,554	27,855,444	3.36

Sources: Texas Department of Health, "Population Data System," Austin, November 15, 1982. (Unpublished Computer Printout.)

Table 4.13

Population Projections: Mexico and United States

Country	1980	1990	2000	Rate
Mexico (a)	70,690,000	97,929,000	130,508,000	3.07
(b)	69,346,900	86,018,700	100,041,400	1.83
United States	220,497,000	237,028,000	248,372,000	0.60

Source: Secretaría programación y Presupuesto, Datos Básicos Sobre la Población de México 1980-2000 (Mexico, D.F.: SPP, 1981).

U.S. Bureau of the Census, *Illustrative Projections of World Populations To the 21st Century*, Special Studies Series, P-23, no. 79, (Washington, D.C.: U.S. Government Printing Office, 1979).

The U.S. Bureau of the Census projected the population figure beside the Mexico (a) and the United States headings. The Secretaría de Programación y Presupuesto projected the Mexico (b) estimates.

Table 4.14

Population Projections for the Mexican Border Region

State	1960	1970	1980	1990	2000
Chihuahua	1,227,000	1,613,000	1,933,856	2,428,000	3,048,000
Coahuila	908,000	1,115,000	1,558,401	2,101,000	2,777,000
Nuevo León	1,079,000	1,695,000	2,463,298	3,721,000	5,622,000
Tamaulipas	1,204,000	1,457,000	1,924,934	2,639,000	3,619,000
Total	4,238,000	5,880,000	7,880,489	10,889,000	15,066,000
Mexico	36,106,000	49,417,000	67,382,581	92,051,000	125,750,000

Sources: Patricia M. Rowe, *Country Demographic Profiles Mexico*, U.S. Bureau of the Census (Washington, D.C.: U.S. Government Printing Office, September 1979).

Secretaría de Programmación Presupuesto, *X Census General de Población y Vivienda, 1980*, Resultados preliminares a Nivel Nacional y por Entidad Federativé (México, D.F.: Secretaría de Programación y Presupuesto, 1981).

Table 4.15

Two Population Projections for Mexican Border Cities

City	1980	1990	2000	Rate
Ciudad Juarez	578,471	694,636	786,339	1.83
		834,129	1,068,903	3.07
Ojinaga	32,068	36,798	44,188	1.83
		41,656	56,624	3.07
Ciudad Acuña	47,330	55,597	66,762	1.83
		62,936	85,552	3.07
Piedras Negras	45,048	79,884	95,926	1.83
		90,430	122,925	3.07
Nuevo Laredo	238,740	286,682	324,529	1.83
		344,252	441,146	3.07
Reynosa	244,913	294,353	332,920	1.83
		353,154	452,552	3.07
Matamoros	242,973	291,765	330,283	1.83
		350,356	448,967	3.07
Total	1,429,543	1,716,616	1,943,236	1.83
		2,061,337	2,641,520	3.07

Source: Edmundo Victoria Mascorro, "Característica del Desarrollo Económico de la Franja Fronteriza Norte de México," *Natural Resources Journal* 22, no. 4 (October 1982): 827.

The 1980 populations of Ojinaga, Ciudad Acuña and Piedras Negras were computed by extrapolating 1970-80 growth rates for states of Chihuahua and Coahuila, 1.81 and 3.54 percent respectively. The growth rates used in this table are based on the estimates by the Secretaría de Programmación Presupuesto (1.83%) and the U.S. Bureau of the Census (3.07%).

Table 5.1

Estimated Water Withdrawals in the Rio Grande/Río Bravo Basin, 1974-1976
(acre-feet)

Source	Texas	Mexico	Total
Surface	1,428,100	3,182,500	4,610,600
Ground	1,209,800	54,000*	1,263,800
Total	2,637,900	3,236,500	5,874,400

Sources: Felipe Oshoa y Asociados, S.C., *Estudio de la Calidad del Agua en la Cuenca del Río Bravo* (México, D.F.: Secretaría de Agricultura y Recursos Hidráulicos, September 1978). (Limited Distribution Document.)

Texas Department of Water Resources, Planning and Development Division, Economics Section, "Fresh Water Use in Texas: 1974 and 1977," Austin, 1982. (Unpublished Computer Printout.)

Figures are rounded to the nearest hundred acre-feet. Texas data (for 1974 only) include groundwater and surface water withdrawals. Mexican figures are for 1974 municipal withdrawals and 1976 irrigation withdrawals only. Mexican municipal water withdrawals were estimated by multiplying daily annual withdrawal rates for 1974 by 365.

* No information is available regarding Mexican groundwater withdrawals for irrigation.

Table 5.2

Estimated Annual Water Withdrawals in the
Rio Grande/Río Bravo Basin by Reach
(acre-feet)

Source	Upper	Middle	Lower
Surface	1,598,000	470,200	2,628,800
Ground	790,700	247,000	30,700
Total	2,388,700	717,200	2,659,500

Sources: Felipe Ochoa y Asociados, S.C., *Estudio de la Calidad del Agua en la Cuenca del Río Bravo*, (México, D.F.: Secretaría de Agricultura y Recursos Hidráulicos, September 1978). (Limited Distribution Document.)

Texas Department of Water Resources, Planning and Development Division, Economics Section, "Fresh Water Use in Texas: 1974 and 1977," Austin, 1982. (Unpublished Computer Printout.)

Texas Department of Water Resources, Planning and Development Division, Economics Section, "1980 Water Use by Category and Source," Austin, 1982. (Unpublished Computer Printout.)

Figures are rounded to the nearest hundred acre-feet. Mexican data include 1974 municipal water withdrawals and 1976 irrigation withdrawals. Texas water withdrawal data represent an average for reported use from 1974, 1977, and 1980.

Table 5.3

**Estimated 1974, 1977, and 1980 Water Withdrawals on the
Texas Side of the Rio Grande/Río Bravo Basin
(acre-feet)**

Reach	1974	1977	1980	Percentage Change 1974 to 1980
Upper	1,106,200	809,400	1,028,500	-7
Middle	414,600	351,100	261,700	-37
Lower	1,117,000	974,500	1,522,100	+36

Sources: Texas Department of Water Resources, Planning and Development Division, Economics Section, "Fresh Water Use in Texas: 1974 and 1977," Austin, 1982. (Unpublished Computer Printout,)

Texas Department of Water Resources, Planning and Development Division, Economics Section, "1980 Texas Water Use by Category and Source," Austin, 1982. (Unpublished Computer Printout.)

Figures are rounded to the nearest hundred acre-feet.

Table 5.4

Estimated Annual Water Withdrawals by Sector and Reach
(acre-feet)

Sector	Upper	Middle	Lower	Total
Irrigation	2,146,500	665,400	2,489,700	5,301,600
Municipal	171,000	38,350	154,800	364,150
Mining	39,300	1,060	1,900	42,260
Livestock	9,750	9,100	10,000	28,850
Manufacturing	12,400	1,800	6,700	20,900
Steam-Electric	9,700	0	5,900	15,600

Sources: Felipe Ochoa y Asociados, S.C., *Estudio de la Calidad del Agua en la Cuenca del Río Bravo* (México D.F.: Secretaría de Agricultura y Recursos Hidráulicos, September 1978). (Limited Distribution Document.)

Texas Department of Water Resources, Planning and Development Division, Economics Section, "Fresh Water Use in Texas: 1974 and 1977," Austin, 1982. (Unpublished Computer Printout.)

Texas Department of Water Resources, Planning and Development Division, Economics Section, "1980 Water Use by Category and Source," Austin, 1982. (Unpublished Computer Printout.)

Figures are rounded to the nearest hundred acre-feet. Mexican data include 1974 municipal water withdrawals only. Texas water withdrawal data represent an average of reported withdrawals from 1974, 1977, and 1980.

Table 5.5

Self-Reported 1981 Municipal and Industrial Return Flows
in the Texas Border Region
(acre-feet)

Source	Upper	Middle	Lower
Municipal	1485.0	216.0	1710.0
Industrial	1474.0	0.6	8407.0
Total	2959.0	216.6	10117.0

Source: Texas Department of Water Resources, "Self-Reporting System: Waste Load Data Report, 1981," Austin, 1982. (Unpublished Computer Printout.)

Table 5.6

Estimated 1974, 1977, and 1980 Groundwater Reliance
in the Texas Border Reaches
(%)

Reach	1974	1977	1980
Upper	75	89	69
Middle	63	61	61
Lower	3	2	1

Sources: Texas Department of Water Resources, Planning and Development Division, Economics Section, "Fresh Water Use in Texas: 1974 and 1977," Austin, 1982. (Unpublished Computer Printout.)

Texas Department of Water Resources, Planning and Development Division, Economics Section, "1980 Water Use by Category and Source," Austin, 1982. (Unpublished Computer Printout.)

Table 5.7

Daily Municipal Water Withdrawals of the Principal Cities
Located on the Mexican Side of the Rio Grande/Rio Bravo Basin

State	City	Inhabitants	Population Served with Potable Water (%)	Type of Water Supply	Daily Withdrawn (cu m)
Chihuahua	Ciudad Juarez	550,000	92	29 Wells	156,718
	Ojinaga	18,000	70	Conchos River	4,320
				2 Wells	6,134
Coahuila	Ciudad Acuña	42,627	77	Rio Grande	6,500
				1 Well	2,764
				1 Gallery	3,000
	Piedras Negras	93,100	72	Rio Grande	21,600
				1 Well	845
				2 Galleries	8,644
Tamaulipas	Nuevo Laredo	175,000	83	Rio Grande	86,400
	Ciudad Guerrero	3,300	N/R	N/R	N/R
	Ciudad Miguel Alemán	14,000	85	Rio Grande	1,850

Table 5.7 (continued)

State	City	Inhabitants	Population Served with Potable Water (%)	Type of Water Supply	Daily Withdrawn (cu m)
	Gustavo Diaz Ordaz	11,297	60	Rio Grande	2,160
	Reynosa	170,000	66	Rio Grande	40,000
	Río Bravo	50,000	50	6 Wells	8,260
	Matamoros	175,000	70	Soliseno Canal	39,000
	Ciudad Valle Hermoso	30,000	67	Santa Rosa Canal	3,888

Source: Felipe Ochoa y Asociados, S.C., *Estudio de la Calidad del Agua en la Cuenca del Río Bravo* (México, D.F.: Secretaria de Agricultura y Recursos Hidráulicos, September 1978). (Limited Distribution Document.)

N/R = not reported.

Table 5.8

Municipal Water Withdrawals for Contiguous Cities
of the Rio Grande/Río Bravo Basin
(acre-feet)

Reach	City	Year	Water Source Ground	Water Source Surface	Total	Population*	Estimated Per Capita Withdrawals
Upper	El Paso	1970	N/A	N/A	74,225	322,261	0.23
		1974	N/A	N/A	80,520	352,303	0.23
		1980	70,564	18,377	88,941	425,259	0.21
	Ciudad Juarez	1974	46,393	0	46,393	550,000	0.08
Middle	Del Rio	1970	N/A	N/A	4,989	21,330	0.23
		1974	N/A	N/A	6,460	23,395	0.27
		1980	339	9,440	9,779	30,034	0.33
	Ciudad Acuña	1974	1,706	1,924	3,630	42,627	0.09
	Eagle Pass	1970	N/A	N/A	2,717	15,364	0.18
		1974	N/A	N/A	2,783	17,309	0.16
		1980	0	4,211	4,211	21,407	0.20
	Piedras Negras	1974	2,809	6,394	9,203	93,100	0.10

Table 5.8 (continued)

Reach	City	Year	Water Source Ground	Water Source Surface	Total	Population*	Estimated Per Capita Withdrawals
Lower	Laredo	1970	N/A	N/A	15,719	69,024	0.23
		1974	N/A	N/A	15,016	73,730	0.20
		1980	0	22,283	22,283	91,449	0.24
	Nuevo Laredo	1974	0	25,577	25,577	175,000	0.15
	McAllen	1970	N/A	N/A	8,525	37,036	0.23
		1974	N/A	N/A	9,933	40,532	0.25
		1980	0	13,027	13,027	66,281	0.20
	Reynosa	1974	0	11,841	11,841	170,000	0.07

Table 5.8 (continued)

Reach	City	Year	Water Source			Population*	Estimated Per Capita Withdrawals
			Ground	Surface	Total		
	Brownsville	1970	N/A	N/A	11,526	52,522	0.22
		1974	N/A	N/A	11,359	57,055	0.20
		1980	0	17,843	17,843	84,997	0.21
	Matamoros	1974	0	11,545	11,545	175,000	0.07

Sources: Felipe Schoa y Asociados, S.C., *Estudio de la Calidad del Agua en la Cuenca del Río Bravo* (México, D.F.: Secretaría de Agricultura y Recursos Hidráulicos, September 1978). (Limited Distribution Document.)

Texas Water Development Board, "Projections of Population and Municipal Water Needs: Series A, Municipal," Austin, 1982. (Unpublished Computer Printout.)

Texas Department of Water Resources, Planning and Development Division, Water Use Technology Unit, "Municipal Water Use for 1980," Austin, 1982. (Unpublished Computer Printout.)

Mexican municipal water withdrawals were estimated by multiplying daily annual water withdrawal rates for 1974 by 365.

*Mexican municipal population data represent an enumeration of 1970 municipal populations as cited in the IX Censo General de Población.

N/A = not available

Table 5.9

1980 Water Withdrawals by Texas Standard Metorpolitan Statistical Areas (SMSAs) (acre-feet)

COUNTY/City	Ground	Surface	Total	Population
Brownsville-Harlingen-San Benito SMSA				
CAMERON				
Bayview	N/A	N/A	N/A	N/A
Brownsville	0	17,843	17,843	84,887
Combes *	0	111	111	1,441
Harlingen	0	9,160	9,160	43,543
Laferia	0	508	508	3,495
Laguna Vista	N/A	N/A	N/A	N/A
Los Fresnos	0	415	415	2,173
Port Isabel	0	2,106	2,106	3,769
Primera	0	162	162	1,380
Rangerville	N/A	N/A	N/A	N/A
Rio Hondo	0	469	469	1,673
San Benito	0	2,956	2,956	17,988
Santa Rosa	0	221	221	1,889
South Padre	N/A	N/A	N/A	N/A
Subtotal	0	33,951	33,951	162,238
El Paso SMSA				
EL PASO				
Anthony	423	0	433	2,640
Clint	42	0	42	1,340
El Paso	70,564	18,377	88,941	425,254
La Isla	N/A	N/A	N/A	N/A
Vinton	N/A	N/A	N/A	N/A
Subtotal	71,039	18,377	89,416	429,234

Table 5.9 (Continued)

1980 Withdrawals by Texas Standard Metropolitan Statistical Areas (SMSAs)

COUNTY/City	Ground	Surface	Total	Population
	Laredo SMSA			
WEBB				
Laredo	0	22,283	22,283	91,449
Subtotal	0	22,283	22,283	91,449
	McAllen-Pharr-Edinburg SMSA			
HIDALGO				
Alamo	0	1,628	1,628	5,831
Donna	0	1,290	1,290	9,952
Edcouch	0	476	476	3,092
Edinburg	0	3,813	3,813	24,075
Elsa	0	829	829	5,061
Hidalgo	355	0	355	2,288
La Joya	0	514	514	2,018
La Villa	0	100	100	1,442
McAllen	0	13,027	13,027	66,281
Mercedes	499	2,273	2,772	11,851
Mission	0	5,332	5,332	22,859
Pharr	0	3,004	3,004	21,381
San Juan	398	1,111	1,509	7,608
Weslaco	0	3,653	3,653	19,331
Subtotal	1,252	37,050	38,302	203,070

Sources: Texas Department of Water Resources, Planning and Development Division, *Water Use, Projected Water Requirements, and Related Data and Information for the Standard Metorpolitan Areas in Texas*, LP-141 (Austin: TDWR, March 1981).

Texas Department of Water Resources, Planning and Development Division, Water Use Technology Unit, "Municipal Water Use for 1980," Austin, 1982. (Unpublished Computer Printout.)

N/A = not available

Table 5.10

Estimated 1974 Municipal Water Withdrawals for Contiguous Border Cities in the Rio Grande/Río Bravo Basin (acre-feet)

Reach	City	Subtotal	Total for Reach
Upper			126,913
	El Paso	80,520	
	Ciudad Juarez	46,393	
Middle			22,076
	Del Rio	6,460	
	Ciudad Acuña	3,630	
	Eagle Pass	2,983	
	Piedras Negras	9,203	
Lower			85,271
	Laredo	15,016	
	Nuevo Laredo	25,577	
	McAllen	9,933	
	Reynosa	11,841	
	Brownsville	11,359	
	Matamoros	11,545	
Total			234,260

Sources: Felipe Ochoa y Asociados, S.C., *Estudio de la Calidad del Agua en la Cuenca del Río Bravo* (México, D.F.: Secretaría de Agricultura y Recursos Hidráulicos, September 1978). (Limited Distribution Document.)

Texas Water Development Board, "Projections of Population and Municipal Water Needs: Series A Municipal 1970 and 1974," Austin, 1982. (Unpublished Computer Printout.)

Yearly municipal withdrawal rates for Mexican border cities were derived by multiplying published daily annual withdrawal rates for 1974 by 365.

Table 5.11

Population Growth Rates of Selected Mexican Border Cities
1960-1976
(thousands)

City	1960	1970	1976	Percentage Change 1960 to 1976
Ciudad Juarez	262.1	407.4	538.2	+ 105
Reynosa	74.1	137.4	200.5	+ 170
Nuevo Laredo	92.6	148.9	199.5	+ 115

Source: Marynka Olizar, *Guide to Mexican Markets: 1976-1977* (México, D.F.: Merchandizing and Marketing, S.A., 1977), p. 27; cited in Stephen paul Mumme, "The United States-Mexico Groundwater Dispute: Domestic Influence on Foreign Policy" (Ph.D. dissertation, University of Arizona, 1982), p. 134.

Table 5.12

Selected Mexican States Ranked According to Annual
Per Capita Income and Percentage of
National Industrial Product, 1965

State	Annual Per Capita Income		Percentage of National Industrial Product
	Pesos	Index	
Nuevo León	11451.0	242.8	10.5
Coahuila	7895.4	138.9	4.9
Tamaulipas	6774.1	119.2	1.4
Chihuahua	4863.4	85.5	1.6
Mexican Averages	5685.3	100.0	100.0

Source: Ifigenia M. de Navarrete, "La Distribución del Ingreso en México Tendencias y Perspectivas," in *El Perfil de México en 1980*, vol. 1 (México D.F.: Sigeo Vienteuno Editores, S.A., 1970), p. 70; cited in Stephen Paul Mumme, "The United States-Mexico Groundwater Dispute: Domestic Influence on Foreign Policy" (Ph.D. dissertation, University of Arizona, 1982), pp. 140-141.

Table 5.13

Industrial Water Withdrawals on the Texas Side of the
Rio Grande/Río Bravo Basin: 1974, 1977, and 1980
(acre-feet)

Reach	1974	1977	1980	Percentage Change 1974 to 1980
Upper	38,900	74,600	70,900	+82
Middle	13,200	12,400	10,000	-24
Lower	19,900	20,600	15,600	-22
Total	72,000	107,600	96,500	+34

Sources: Texas Department of Water Resources, Planning and Development Division, Economics Section, "Fresh Water Use in Texas: 1974 and 1977," Austin, 1982. (Unpublished Computer Printout.)

Texas Department of Water Resources, Planning and Development Division, Economics Section, "1980 Texas Water Use by Category and Source," Austin, 1982. (Unpublished Computer Printout.)

Figures are rounded to the nearest hundred acre-feet.

Table 5.14

Fuels Mining Water Withdrawals on the Texas Side of the
Rio Grande/Río Bravo Basin
1974 and 1977
(acre-feet)

Reach	1974	1977	Percentage Change 1974 to 1977
Upper	4,980	4,110	-18
Middle	60	10	-77
Lower	440	1,070	+144
Total	5,480	5,190	-5

Source: Texas Department of Water Resources, Planning and Development Division, Economics Section, "Fresh Water Use in Texas: 1974 and 1977," Austin, 1982. (Unpublished Computer Printout.)

Figures are rounded to the nearest ten acre-feet.

Table 5.15

**Water Withdrawals for Mining on the Texas Side of
the Rio Grande/Río Bravo Basin: 1974 and 1977
(acre-feet)**

Reach	Mining	1974	1977	Percentage Change 1974 to 1977
Upper	Metallic	3,500	0	-100
	Nonmetallic	7,630	47,110	+518
Middle	Metallic	0	0	--
	Nonmetallic	700	1,070	+53
Lower	Metallic	0	260	--
	Nonmetallic	1,210	1,490	+23
Total	Metallic	3,500	260	-93
	Nonmetallic	9,530	49,670	+421

Source: Texas Department of Water Resources, Planning and Development Division, Economics Section, "Fresh Water Use in Texas: 1974 and 1977," Austin, 1982. (Unpublished Computer Printout.)

Figures are rounded to the nearest ten acre-feet.

Table 5.16

1974, 1977, and 1980 Livestock Water Withdrawals on the Texas Side of the Rio Grande/Río Bravo Basin
(acre-feet)

Reach	1974	1977	1980	Percentage Change 1974 to 1980
Upper	9,610	9,290	10,350	+8
Middle	10,230	9,360	7,440	-27
Lower	10,450	10,740	9,100	-13

Sources: Texas Department of Water Resources, Planning and Development Division, Economics Section, "Fresh Water Use in Texas: 1974 and 1977," Austin, 1982. (Unpublished Computer Printout.)

Texas Department of Water Resources, Planning and Development Division, Economics Section, "1980 Texas Water Use by Category and Source," Austin, 1982 (Unpublished Computer Printout.)

Figures are rounded to the nearest ten acre-feet.

Table 5.17

**1974, 1977, and 1980 Manufacturing Water Withdrawals on the
Texas Side of the Rio Grande/Río Bravo Basin
(acre-feet)**

Reach	1974	1977	1980	Percentage Change 1974 to 1980
Upper	13,270	14,160	9,780	-26
Middle	2,180	1,970	1,220	-44
Lower	7,790	7,000	5,250	-33

Sources: Texas Department of Water Resources, Planning and Development Division, Water Use Technology Unit, "Manufacturing Water Demand: 1974," Austin, 1982. (Unpublished Computer Printout.)

Texas Department of Water Resources, Planning and Development Division, Water Use Technology Unit, "1977 County Manufacturing Water Use," Austin, 1982. (Unpublished Computer Printout.)

Texas Department of Water Resources, Planning and Development Division, Water Use Technology Unit, "Reported Manufacturing Water Use: 1980," Austin, 1982. (Unpublished Computer Printout.)

Figures are rounded to the nearest ten acre-feet.

Table 5.18

Industrial Water Withdrawal Distribution in Texas, 1971

SIC Code	Manufacturing Sector (standard industrial classifications)	Percentage of Total Withdrawals for Different Water Uses			
		Process	Condensing Cooling and Air Conditioning	Boiler Feed	Other
20	Food and Kindred Products*	44	42	9	5
21	Tobacco Products**	na	na	na	na
22	Textile Mill Products	84	8	3	5
23	Apparel and Related Products	0	8	16	76
24	Limber and Wood Products	59	8	27	6
25	Furniture and Fixtures	62	4	11	33
26	Paper and Allied Products*	87	10	2	1
27	Printing and Publishing	4	59	5	32
28	Chemical and Allied Products*	21	54	20	7
29	Petroleum Refining*	14	53	26	7
30	Rubber and Miscellaneous Plastics	19	51	10	20

Table 5.18 (continued)

		Percentage of Total Withdrawals for Different Water Uses			
		Process	Condensing Cooling and Air Conditioning	Boiler Feed	Other
SIC Code	Manufacturing Sector (standard industrial classification)				
31	Leather and Leather Products	17	2	1	80
32	Stone Clay and Glass Products	18	78	1	3
33	Primary Metals	30	32	12	26
34	Fabricated Metal Products	40	28	18	14
35	Machinery except Electrical	13	45	7	35

Table 5.18 (continued)

SIC Code	Manufacturing Sector (standard industrial classification)	Percentage of Total Withdrawals for Different Water Uses			
		Process	Condensing Cooling and Air Conditioning	Boiler Feed	Other
36	Electrical Machinery	41	17	1	41
37	Transportation Equipment	6	87	1	6
38	Mechanical and Scientific Instruments	25	39	2	34
39	Miscellaneous Industries	35	23	4	38
	All Manufacturing Industries	32	46	16	6

Source: Texas Water Development Board, "Continuing Water Resources Planning and Development for Texas: Phase I," Volume I, Austin, May 1977, p. II-74. (Draft.) *Indicates the five industrial sectors that together account for over 85 percent of total industrial water withdrawals in Texas. **The tobacco industry (SIC 21) is not found in Texas. na = not applicable.

Table 5.19

**1974, 1977, and 1980 Irrigation Water Withdrawals on the
Texas Side of the Rio Grande/Río Bravo Basin
(acre-feet)**

Reach	1974	1977	1980	Percentage Change 1974 to 1980
Upper	859,000	548,600	731,900	-15
Middle	381,600	359,600	307,000	-20
Lower	1,100,000	852,800	1,400,000	+27

Sources: Texas Department of Water Resources, Planning and Development Division, Economics Section, "Fresh Water Use in Texas: 1974 and 1980," Austin, 1982. (Unpublished Computer Printout.)

Texas Department of Water Resources, Planning and Development Division, Economics Section, "1980 Texas Water Use by Category and Source," Austin, 1982. (Unpublished Computer Printout.)

Figures are rounded to the nearest hundred acre-feet.

Table 5.20

1966, 1968, and 1976 Irrigation Water Withdrawals on the Mexican Side of the Rio Grande/Río Bravo Basin
(acre-feet)

Reach	1966	1968	1976	Percentage Change 1966 to 1976
Upper*	523,000	1,024,900	1,433,500	+174
Middle	77,200	181,900	316,000	+309
Lower	664,000	742,100	1,372,100	+106
Total	1,264,200	1,948,900	3,121,600	+147

Sources: Felipe Ochoa y Asociados, *Estudia de la Calidad del Agua en la Cuenca del Río Bravo* (México, D.F.: Secretaría de Agricultura y Recursos Hidráulicos, September 1978). (Limited Distribution Document.)

Secretaría de Recursos Hidráulicos, *Características de los Distritos de Reigo: Zonas Pacifico Norte, Norte Centro, y Noreste*, Tomo I (México, D.F.: Secretaría de Recursos Hidráulicos, 1969).

Secretaría de Recursos Hidráulicos, *Características de los Distritos de Reigo: Zonas Pacifico Norte, Norte Centro, y Noreste*, Tomo I (México, D.F.: Secretaría de Recursos Hidráulicos, 1969).

Figures are rounded to the nearest hundred acre-feet.

*The upper reach does not include any data for the Bajo Río Conchos irrigation district in either 1966 or 1968.

Table 5.21

Irrigation Water Withdrawals along the Mexican Side of the Rio Grande/Rio Bravo Basin: 1966, 1968, and 1976

Irrigation District	Main Stream	Irrigation Areas (ha)			Total Volume Used (cu m)		
		1966	1968	1976	1966	1968	1976
Bajo Rio Conchos	Conchos River	N/A	N/A	2,100	N/A	N/A	160,846
Ciudad Delicius	Conchos River	37,642	68,400	58,903	514,728	1,087,121	1,401,891
	San Pedro River						
Valle de Juarez	Rio Grande River Juarez Wastewaters	14,067	14,715	14,500	130,197	176,600	204,240
Palestina	San Diego River	4,898	4,567	10,534	37,168	43,485	187,400
Don Martin	Salado River	6,431	14,772	16,000	53,900	177,664	187,680
Acuña Falcon	Bravo River	7,535	5,377	4,651	41,278	23,801	29,301
Bajo Rio Bravo	Bravo River	85,287	149,389	206,718	548,534	626,705	1,093,538

Table 5.21 (continued)

Irrigation District	Main Stream	Irrigation Areas (ha)			Total Volume Used (cu m)		
		1966	1968	1976	1966	1968	1976
Alto Río San Juan	San Juan River	1,041	867	1,718	4,131	3,141	14,569
Las Lajas	San Juan River	4,443	2,564	3,499	11,838	6,774	30,489
Bajo Río San Juan	San Juan River	35,432	58,948	75,000	217,260	257,785	538,500
Subtotal		197,776	319,609	403,594	1,559,034	2,403,076	3,848,954

Sources: Felipe Ochoa y Asociados, S.C., *Estudio de la Calidad del Agua en la Cuenca del Río Bravo* (México, D.F.: Secretaría de Agricultura y Recursos Hidráulicos, September 1978). (Limited Distribution Document.)

Secretaría de Recursos Hidráulicos, *Características de los Distritos de Reigo: Zonas Pacifico Norte, Norte Centro, y Noreste, Tomo I* (México, D.F.: SRH, 1967).

Secretaría de Recursos Hidráulicos, *Características de los Distritos de Reigo: Zonas Pacifico Norte, Norte Centro, y Noreste, Tomo I* (México, D.F.: SRH, 1969).

N/A = not available

Table 5.22

1974, 1977, and 1980 Steam-Electric Water Withdrawals on the Texas Side of the Rio Grande/Río Bravo Basin (acre-feet)

Reach	1974	1977	1980	Percentage Change 1974 to 1980
Upper	7,750	8,300	13,000	+68
Middle	0	0	0	--
Lower	6,240	4,490	7,130	+14

Sources: Texas Department of Water Resources, Planning and Development Division, Economics Section, "Fresh Water Use in Texas: 1974 and 1977," Austin, 1982. (Unpublished Computer Printout.)

Texas Department of Water Resources, Planning and Development Division, Economics Section, "1980 Texas Water Use by Category and Source," Austin, 1982. (Unpublished Computer Printout.)

Figures are rounded to the nearest ten acre-feet.

Table 6.1

Water Disinfection Failures in
Texas Public Water Systems, 1981-1982
(%)

Violation	Community System	Noncommunity System
Lack of or inadequate disinfection	20	26
Failure to maintain 0.5 mg/l chlorine residual in far reaches of distribution system	30	42

Source: Compiled by Deborah Sagen from data obtained from 1981-1982 annual water system compliance surveys of the Texas Department of Health.

Table 6.2

**Violations of Maximum Contaminant Levels in
Texas Public Water Systems, 1981-1982
(%)**

Contaminant	Community Systems	Noncommunity Systems
Chlorides	4	13
Fecal Coliform	6	14
Fluoride	12	*
Nitrate	*	5
Sulphate	*	14
Total Dissolved Solids	4	14
Turbidity	12	*

Source: Compiled by Deborah Sagen from 1981 and 1982 annual water system compliance surveys and other information in files of the Texas Department of Health.

*Indicates that fewer than four percent of the systems committed these violations.

Table 7.1

Residential Plumbing Facilities in
Selected Texas Counties, 1970
(%)

County	Homes Served by Public Sewer, Septic Tank, or Cesspool	Homes Lacking Some or All Plumbing
Cameron	87.9	19.1
Hidalgo	80.5	25.1
Webb	91.3	16.7
Maverick	83.7	7.3
Val Verde	94.3	10.5
El Paso	98.1	7.9

Sources: U.S. Department of Commerce, Bureau of the Census, *1970 Census of the Population* (Washington, D.C.: U.S. Government Printing Office, 1972), Tables 60, 62.

U.S. Department of Commerce, *City County Data Book, 1970* (Washington, D.C.: U.S. Government Printing Office, 1970), Items 65-80.

Table 7.2

Residential Plumbing Facilities in
Selected Mexican Cities, 1970
(%)

Cities	Inside Piped Water without Sewerage	Outside Piped Water without Sewerage	No Piped Water without Sewerage
Matamoros	4.6	10.5	26.8
Reynosa	5.6	9.4	21.4
Nuevo Laredo	3.8	11.7	10.6
Piedras Negras	8.5	14.1	9.5
Acuña	8.5	16.1	25.6
Juarez	4.2	10.8	12.9

Source: Secretaría de Programación y Presupuesto, Coordinación General de los Servicios Nacionales de Estadística, Geografia e Informática, *Censo General de Población, 1970* (México, D.F.: SPP, 1972), Tables 33-35.

Table 7.3

Accessibility to Public Drainage or Septic Tanks
in Mexico, 1980
(%)

State	Homes without Public Drainage or Septic Tanks	Homes with Public Drainage or Septic Tanks
Tamaulipas	49.5	50.5
Coahuila	43.6	56.4
Chihuahua	43.2	56.8

Source: Secretaría de Programación y Presupuesto, *X Censo General de Población y Vivienda, 1980* (México, D.F.: SPP, 1980), Table 21, p. 78. (Preliminary Census Results.)

Table 8.1

Estimated Reduction of Water-Related Diseases
by Water Improvements

Category	Disease	Reduction of Incidence (%)
I	Cholera	90
I	Typhoid	80
I	Leptospirosis	80
I	Tularemia	40
I	Paratyphoid	40
I	Infective hepatitis	10
I	Some enteroviruses	10
I, II	Bacillary dysentery	50
I, II	Amoebic dysentery	50
I, II	Gastroenteritis	50
II	Skin sepsis and ulcers	50
II	Trachoma	60
II	Conjuntivitis	70
II	Scabies	80
II	Yaws	70
II	Leprosy	50
II	Tinea	50
II	Louse-borne fevers	40
II	Diarrhoeal diseases	50
II	Ascariasis	40
III	Schistosomiasis	60
III	Guinea worm	100
IV	Sleeping sickness	80
IV	Onchocerciasis	20
IV	Yellow Fever	10

Source: David J. Bradley, "Health Aspects of Water Supplies in Tropical Countries," in *Water, Wastes and Health in Hot Climates*, ed. Richard Feachem, Michael McGarry, and Duncan Mara (New York: John Wiley & Sons, 1977), p. 9.

Table 8.2

**Mechanisms of Water-Related Disease Transmission
and Preventive Strategies**

Transmission Mechanism	Preventive Strategy
Waterborne	Improve water quality Prevent use of polluted sources
Water-washed	Improve water quantity Improve water accessibility Improve hygiene
Water-based	Decrease need for water contact Control intermediate host populations Improve quality
Water-related insect vector	Improve surface-water management Destroy breeding sites of insects Decrease need to visit breeding sites

Source: David J. Bradley, "Health Aspects of Water Supplies in Tropical Countries," in *Water, Wastes and Health in Hot Climates*, ed. Richard Feachem, Michael McGarry, and Duncan Mara (New York: John Wiley & Sons, 1977), p. 7.

Table 8.3

Water Facilities for Selected Texas Counties, 1980

County	Percentage Lacking Complete Plumbing		Total Occupied Housing Units
	Persons	Housing Units	
Cameron	9.7	8.3	58,418
El Paso	2.7	3.2	140,806
Hidalgo	11.6	9.9	75,816
Maverick	11.2	12.4	7,583
Val Verde	2.4	3.4	10,355
Webb	5.8	6.8	25,896

Source: Calculations by Alfonso Ortiz based on U.S. Department of Commerce, Bureau of Census, *Characteristics of Housing Units*, chapter A, vol. 1, part 45 (Washington, D.C.: U.S. Government Printing Office, 1982).

Table 8.4

Water Facilities in Mexico by State, 1980

State	Percentage of Housing Lacking Infrastructure		Total Housing units
	Complete Plumbing	Public Drainage	
Chihuahua	22.5	43.2	378,736
Coahuila	14.5	43.6	288,114
Nuevo León	11.2	23.5	444,164
Tamaulipas	26.5	49.5	378,020

Source: Calculations by Alfonso Ortiz Nuñez based on Coordinación General de los Servicios Nacionales de Estadística, Geografia e Informática, *X Censo General de Población y Vivienda, 1980: Resultados Preliminares a Nivel Nacional y por Entidad Federativa*, Cuadro 21 (México: Secretaría de Programación y Presupuesto, 1980), pp. 86-89.

Table 8.5

Reported Rates of Enteric Diseases
in Texas Municipalities, 1980
(number of cases per 100,000 population)

Municipality	County	Amebiasis	Infectious Hepatitis	Salmonellosis	Shigellosis	Typhoid
Brownsville	Cameron	25.9	8.2	29.4	63.5	N/A
McAllen	Hidalgo	22.4	6.0	20.9	7.5	N/A
Laredo	Webb	N/A	37.2	37.2	21.9	8.7
Eagle Pass	Maverick	N/A	N/A	4.7	N/A	N/A
Del Rio	Val Verde	N/A	3.3	6.7	N/A	N/A

Source: Calculations by Alfonso Ortiz Nuñez from several unpublished sources of the Texas Department of Health and local health departments.

N/A = not available

Table 8.6

Reported Rates of Enteric Diseases in Texas Counties, 1980
(number of cases per 100,000 population)

Counties	Amebiasis	Infectious Hepatitis	Salmonellosis	Shigellosis	Typhoid
Cameron	11.9	7.2	21.0	33.4	0.0
El Paso	2.0	37.5	19.8	28.3	1.0
Hidalgo	19.8	2.8	24.7	28.6	2.5
Maverick	0.0	0.0	0.0	0.0	0.0
Val Verde	0.0	0.3	0.0	0.0	0.0
Webb	0.0	34.3	55.4	0.0	8.1

Source: Texas Department of Health, Bureau of Communicable Disease Services, *Reported Morbidity and Mortality* (Austin: Division of Public Health Education, 1981).

Table 8.7

Reported Rates of Enteric Diseases in Texas, 1976-1980
(number of cases per 100,000 population)

Year	Amebiasis	Salmonellosis	Shigellosis	Typhoid	Infectious Hepatitis
1976	1.1	7.3	10.4	0.1	13.9
1977	1.7	8.1	12.2	0.2	16.2
1978	1.6	9.2	14.3	0.3	20.7
1979	2.2	16.4	17.2	0.5	24.6
1980	2.5	17.3	15.2	0.5	20.9
U.S. rates 1980	2.3	14.9	8.4	N/A	12.8

Source: Texas Department of Health, Bureau of Communicable Disease Services, *Texas Morbidity This Week* (Austin: Division of Public Health Education, 1981).

N/A = not available

Table 8.8

Reported Rates of Enteric Diseases
in Mexican States, 1976-1979
(number of cases per 100,000 population)

States	Years	Amebiasis	Salmonellosis	Shigellosis	Typhoid	Infectious Hepatitis
Tamaulipas	1976	55.0	3.8	0.5	3.3	3.4
	1977	77.1	2.4	1.9	2.1	5.5
	1978	802.5	3.5	26.2	209.6	6.8
	1979	703.4	N/A	24.0	10.2	50.8
Coahuila	1976	975.8	180.9	15.3	5.7	8.8
	1977	894.3	239.2	11.4	11.8	32.8
	1978	1,292.8	241.2	12.3	282.8	49.4
	1979	1,283.9	N/A	38.4	24.9	46.4

Table 8.8 (continued)

States	Years	Amebiasis	Salmonellosis	Shigellosis	Typhoid	Infectious Hepatitis
Chihuahua	1976	276.7	55.2	18.2	4.8	24.1
	1977	273.0	113.9	7.5	6.7	34.3
	1978	480.7	67.6	7.4	142.9	31.2
	1979	572.2	N/A	20.0	3.4	20.4
Nuevo León	1976	592.7	5.9	6.7	12.2	6.7
	1977	707.1	4.2	16.4	9.5	10.0
	1978	616.2	3.5	15.9	1.6	3.4
	1979	N/A	N/A	N/A	N/A	30.0

Source: Secretaría de Salubridad y Asistencia, Unidad de Información, "Reporte de Enfermedades Comunicables, México 1976-1979," México, D.F., 1979. (Unpublished Data.)

N/A = not available

Table 8.9

Mortality Rates for Infectious Intestinal Diseases
in Six Texas Counties
(number of deaths per 100,000 population)

County	1970	1976	1978	1980
Cameron	7.1	0.0	0.5	0.0
El Paso	1.4	2.0	0.8	0.0
Hidalgo	8.2	2.8	1.5	0.0
Maverick	0.0	0.0	3.4	0.0
Val Verde	3.6	0.0	5.8	2.8
Webb	4.1	0.0	2.1	0.0

Source: Calculations by Alfonso Ortiz Nuñez based on death records in computer tapes provided by the Texas Department of Health.

Table 8.10

Deaths from Selected Causes in Texas Counties, of Residence, 1980-1981

Counties	Years	Shigellosis and Amebiasis	Viral Hepatitis	Other Intestinal infectious	Other infectious and Parasitic Diseases
Cameron	1980	0	1	0	0
	1981	0	1	0	1
El Paso	1980	0	9	0	2
	1981	0	0	2	12
Hidalgo	1980	0	3	0	2
	1981	0	1	0	6
Maverick	1980	0	0	0	0
	1981	0	0	0	0
Val Verde	1980	0	0	1	1
	1981	0	0	0	1
Webb	1980	0	3	0	0
	1981	0	0	0	1

Sources: Texas Department of Health, Bureau of Vital Statistics, *Texas Vital Statistics 1980* (Austin: Division of Public Health Education, 1980), pp. 32–44.

Texas Department of Health, Bureau of Vital Statistics, *Texas Vital Statistics 1981* (Austin: Division of Public Health Education, 1981), pp. 32–44.

Table 8.11

Enteric Disease Mortality Rates in Children 0-4 Years
in Mexican Municipios, 1978
(number of cases per 100,000 population)

Municipio	State	Amebiasis Shigellosis	Infectious Hepatitis	Salmonellosis	Enteritis	Typhoid
Matamoros	Tamaulipas	2.1	N/A	47.9	19.2	1.3
Reynosa	Tamaulipas	1.7	1.3	0.4	35.4	N/A
Nuevo Laredo	Tamaulipas	0.4	N/A	N/A	25.9	N/A
Piedras Negras	Coahuila	2.1	N/A	N/A	38.6	N/A
Ciudad Juarez	Chihuahua	2.5	0.6	1.3	60.3	0.3

Source: Dirección General de Estadística, "Defunciones Generales por Entidad Federativas y Municipios de Residencia," México, D.F., 1979. (Unpublished Data.)

N/A = not available

Table 9.1

Groups of Counties in Rio Grande/Río Bravo Basin

Counties by Reach of the River

Lower Reach	Middle Reach	Upper Reach
Brooks	Dimmit	Brewster
Cameron	Edwards	Crane
Hidalgo	Kinney	Crocket
Jim Hogg	La Salle	Culberson
Kennedy	Maverick	El Paso
Starr	Real	Hudspeth
Webb	Uvalde	Jeff Davis
Willacy	Val Verde	Loving
Zapata	Zavala	Pecos
		Presidio
		Terrell
		Upton
		Ward
		Winkler

Counties in Irrigation Areas

Trans Pecos	Winter Garden	Rio Grande Valley
Culberson	Atascosa	Cameron
El Paso	Bexar	Hidalgo
Hudspeth	Frio	Starr
Jeff Davis	Kinney	Willacy
Pecos	La Salle	Zapata
Presidio	Medina	
Reeves	Uvalde	
Ward	Wilson	

Source: Texas Department of Water Resources, "Water Planning Projections for Texas--1980-2030," Austin, 1982, p. 5. (Draft.)

Table 9.2

Total Industrial Water Withdrawal Projections
(acre-feet)

Reach	Estimate	1980	1990	2000	2010	2020	2030	Change 1980 to 2030 Percent	Change 1980 to 2030 CAGR
Lower	low	5,252	6,261	7,747	9,308	11,245	13,598	158.9	1.92
	high	5,252	7,247	9,247	12,670	16,114	20,237	285.3	2.73
Middle	low	1,212	1,444	1,781	2,140	2,583	3,113	156.9	1.90
	high	1,212	1,627	2,201	2,796	3,534	4,414	264.2	2.62
Upper	low	9,789	9,487	9,977	10,458	12,254	14,369	46.8	0.77
	high	9,789	12,753	15,383	17,704	20,662	24,257	147.8	1.83
Total	low	16,253	17,192	19,305	21,906	26,082	31,080	91.2	1.31
	high	16,253	21,627	27,471	32,927	40,310	48,908	253.0	2.23

Source: Texas Department of Water Resources, "Water Planning Projections for Texas–1980-2030," Austin, 1982, pp. 103-233. (Draft.)

CAGR = continuous annual growth rate.

Table 9.3

Water-Use Projections for Steam-Electric Power
(acre-feet)

Basin	1980	1990	2000	2010	2020	2030
Nueces-Rio Grande	5,593	5,593	5,593	10,394	15,196	20,000
Rio Grande	11,418	11,418	27,180	30,688	34,198	37,712

Source: Texas Department of Water Resources, "Water Planning Projections for Texas--1980-2030," Austin, 1982, p. 376. (Draft.)

Table 9.4

Irrigation and Livestock Water Withdrawal Projections

A. Irrigation Withdrawals
(thousands of acre-feet)

Region	1980	1990	2000	2010	2020	2030
Rio Grande Valley	1,070	1,060	1,060	1,060	1,140	1,140
Winter Gardens	610	280	280	300	300	530
Trans Pecos	700	1,170	1,170	1,170	1,170	1,170

B. Livestock Withdrawals
(acre-feet)

Region	1980	1990	2000	2010	2020	2030
Rio Grande Valley	3,303	3,920	4,540	4,540	4,540	4,540
Winter Gardens	15,850	18,530	21,215	21,215	21,215	21,215
Trans Pecos	6,861	8,020	9,235	9,235	9,235	9,235

Source: Texas Department of Water Resources, "Water Planning Projections for Texas--1980-2030," Austin, 1982, pp. 364-72. (Draft.)

Table 9.5

Results of Statewide Per-Capita Regression Equation

Variable	Beta	Standard Error	Significance
X(2)	137.80	11.27	.0001
X(4)	-89.22	5.28	.0001
X(1)	-58.46	4.92	.0001
X(3)	35.58	4.82	.0001

$R^2 = .39$ degrees of freedom   = 2,075

$R^2 = .62$ F ratio of equation = 54.432

Mean per-capita water consumption is 143 gallons per-capita-per-day.

Source: Texas Department of Water Resources, Planning and Development Division, Water Requirements and Uses Section, "Methodology for Projecting Municipal Water Needs," Austin, 1976, p. 23. (Draft.)

Table 9.6

Total Municipal Water Withdrawal Projections
(acre-feet)

Reach	Estimate	1980	1990	2000	2010	2020	2030	Change, 1980 to 2030 Percent	CAGR
Lower	low	125,197	159,622	204,859	244,138	290,341	334,692	167.3	1.99
	high	125,197	233,274	319,194	414,793	537,547	676,483	440.3	3.43
Middle	low	30,975	33,975	42,843	51,230	59,306	66,138	113.5	1.53
	high	30,975	47,741	64,141	83,499	105,106	127,964	313.1	2.88
Upper	low	126,050	147,917	171,129	193,271	218,510	239,865	90.2	1.30
	high	126,050	209,344	266,549	315,929	386,468	518,330	311.2	2.87
Total	low	282,222	341,414	418,831	488,639	568,157	640,805	127.1	1.65
	high	282,222	490,359	649,884	814,221	1,029,121	1,322,779	368.7	3.14

Source: Texas Department of Water Resources, "Water Planning Projections for Texas--1980-2030," Austin, 1982, pp. 103-233. (Draft.)

CAGR - continuous annual growth rate

Table 9.7

Rio Grande/Río Bravo Total Basin Withdrawals
(acre-feet)

Sector	Estimate	1980	1990	2000	2010	2020	2030	Change, 1980 to 2030 Percent	CAGR
Irrigation	low	2,380,000	2,510,000	2,510,000	2,510,000	2,530,000	2,530,000	6.3	0.12
	high	2,380,000	2,510,000	2,510,000	2,530,000	2,610,000	2,840,000	19.3	0.35
Municipal	low	282,222	341,414	418,831	488,639	568,157	640,805	127.1	1.65
	high	282,222	470,359	649,884	814,221	1,029,121	1,322,779	368.7	3.14
Industrial	low	16,253	17,192	19,305	21,906	26,082	31,080	91.2	1.31
	high	16,253	21,627	27,471	32,927	40,310	48,708	200.9	2.23
Livestock		26,014	30,470	34,990	34,990	34,990	34,990	34.5	0.59
Total	low	2,704,489	2,899,076	2,983,126	3,055,535	3,159,229	3,236,875	19.7	0.36
	high	2,704,489	3,032,456	3,222,345	3,412,138	3,714,421	4,246,677	57.0	0.91

Source: Texas Department of Water Resources, "Water Planning Projections for Texas--1980-2030," Austin, 1982, (Draft.)

CAGR = continuous annual growth rate

Glossary

Acre-foot: a measure of water contained in an area one foot deep and one acre square (equal to 325,851 gallons).

Biochemical oxygen demand (BOD): waste organic matter is stabilized through its decomposition by living bacterial organisms that require oxygen. BOD is a measure of the requirement for oxygen as this matter decomposes and is an index of the degree of organic pollution.

Boron (B): a minor constituent of rocks and water. Excessive boron content will make water unsuitable for irrigation and will be toxic for some crops.

Calcium (Ca) and Magnesium (Mg): inorganic elements dissolved from almost all rocks. Highest concentrations are found in water that has been in contact with limestone or gypsum. These two materials cause hardness in water.

Carbonate (CO3) and Bicarbonate (HCO3): chemical constituents of water that serve as an indication of alkalinity. Carbonate occurs where water is treated with lime. Bicarbonate results when carbon dioxide in water reacts with carbonate rocks.

Chloride (Cl): an element that dissolves from rock materials. The following contribute to an increase in chloride levels: aridity, return drainage from irrigation, sewage, drainage from oil wells, salt springs, and industrial waste. Increased levels of chloride will heighten the corrosive effects of water. Chloride, combined with sodium, causes a salty taste.

Community water system: a drinking-water system that serves at least 15 connections continuously or that regularly serves at least 25 year-round residents.

Consumption: a portion of water withdrawn for offstream use that is not returned to a surface or groundwater source, having been transformed through evapotranspiration or incorporation into products, crops, livestock, or humans.

Discharge: volume of fluid and suspended sediment that passes a given point within a given period of time.

Discharge-weighted average: a measure determined by multiplying discharge for a sampling period by concentrations of individual constituents and dividing by the sum of the discharges. Approximates the composition of water after thorough mixing.

Dissolved oxygen (O2) (DO): a measure of gaseous oxygen in water. The level of DO is affected both by the demand for oxygen placed on water (BOD) and the rate of restoration of oxygen in water. The DO concentration also is affected by the rate of flow, discharge points, temperature, and amount of plant life in the water. Low levels of DO are detrimental to fish life and cause water to become more corrosive.

Effluent: wastewater discharged by an industry or municipality.

Effluent-limited: TDWR classification for stream segments that meet present water quality standards or which will meet standards following incorporation of best practical treatment for industries or secondary treatment for cities.

Evapotranspiration: evaporation and plant consumption of water. Water that evapotranspires is not recharged to surface or groundwater.

Fecal coliform: bacteria of the coliform group originating from the intestines of warm-blooded animals. Occurrence signifies contamination by human wastes; higher levels indicate a potential health hazard.

Fluoride (F): an inorganic chemical that dissolves from most rocks and soils and is often added to water for municipal supplies. Increased levels of fluoride reduce tooth decay up to a point, then may cause mottling of teeth.

Gaging station: a site where records of flow, discharge, and water contents are taken. Surface water quality measurements are often taken at or near gaging stations.

Groundwater: water that exists beneath the land surface.

Groundwater mining: withdrawals from an aquifer at rates in excess of net recharge. The problem becomes serious when this practice continues on a sustained basis over time; groundwater tables decline, making the pumping of water more expensive. At some point aquifer compaction may occur, adversely affecting storage capacity and transmissivity. Quality may also be threatened by salt-water intrusion.

Hardness: a measure of the ability of water to form lather when combined with soap. Soft water produces more lather. Harder water is suited for agricultural irrigation but is detrimental to fabrics, pipes, and boilers.

Iron (Fe): an element dissolved from rocks and soils. High levels of iron in water cause discoloration, straining, and bad taste, thus affecting domestic and industrial use.

Leaching: the process of removal of salts and alkali from soils that occurs when water percolates through the soil.

Manganese (Mn): an element that acts similarly to iron. May often be found in large reservoirs where dissolved from bottom muds. Levels above 2 ppm usually cause staining of fabrics and discoloration of water.

Nitrate (NO3): a final oxidation product caused by decaying organic matter, sewage, and fertilizer. High nitrate levels in drinking water elevate risks of infant methhemoglobinemia.

Noncommunity water system: a TDWR term for any public water system not classified as a community water system.

Nonpoint source: source of potential pollution that is difficult to pinpoint or measure. Common examples include runoff from agricultural areas, forest lands, mining or construction, and urban storms.

pH: a measure of alkalinity or acidity in water. Lower pH indicates acidity and is caused by salts, acids, and carbon dioxide. Higher pH levels may be attributed to carbonates, phosphates, silicates, and borates. Extremely high levels are corrosive.

Phosphate (PO4): a natural nutrient. Excess phosphate may be caused by oxidation of organic matter, water treatment, fertilizer, detergents, and industrial and municipal sewage. High concentrations can be detrimental to fish life, or enhance nuisance algal growth and eutrophication.

Point source: specific discharge source which can be readily identified. Examples include industrial plants and municipal sewage treatment plants.

Public water system: a Texas Department of Health term for a system which provides piped water for human consumption. Such a system has at least 15 service connections or regularly serves an average of at least 25 individuals daily at least 60 days per year. Includes any collection, treatment, storage, and distribution facilities under control of the operator and used in connection with the system. Can also include any collection or pre-treatment storage facilities not under control of the operator and used in connection with the system.

Recharge: refers to water added to the zone of saturation of an aquifer.

Return flow: the portion of withdrawn water that is not consumed by evapotranspiration or incorporation into products, crops, livestock, or humans but returns to its source or to another body of water.

Salts: includes compounds of certain cations (calcium, sodium, potassium, magnesium) and anions (carbonate, bicarbonate, chloride, sulfate, and nitrate). High levels of salts affect plant growth adversely by physically preventing water uptake, directly poisoning plants, or changing soil structure. Indicators of salinity include total dissolved solids (TDS), sodium, specific conductance, and boron.

Sediment: solid material that originates mostly from disintegrated rocks and is transported by, suspended in, or deposited from water. It includes chemical and biochemical precipitates and decomposed organic material. The quantity, characteristics and cause of the occurrence of sediment are influenced by such environmental factors as degree of slope of basin, length of slope, soil characteristics, land usage, and quantity and intensity of rainfall.

Silica ($SiO2$): a material that occurs in sand, quartz, feldspar, and other minerals and can be dissolved from rocks. High silica levels affect the usefulness of water by contributing to boiler scales and turbine blade deposits.

Sodium (Na): an element dissolved from rocks and soils, oil-field brines, water, industrial brines, and sewage. Increased levels will adversely affect irrigation and industrial water uses, and will cause a salty taste in drinking water.

Sodium adsorption ratio: an expression of the reaction of sodium ions with a specific soil. This provides an index of sodium or alkalinity hazard in the soil, especially used for irrigation purposes.

Specific conductance or conductivity: a measure of the ability of water to conduct an electrical current. The measure is used as a surrogate for the amount of dissolved solids or salt content in the water.

Temperature: a measure of heat content. High temperatures restrict the species of fish that can survive in a body of water. Warmer waters contain a reduced capacity to hold dissolved oxygen. At warmer temperatures, oxygen is used more rapidly by organic wastes, lowering oxygen content.

Total dissolved solids (TDS): a measure of dissolved materials in water that indicates salinity. For many purposes, TDS content is a major limitation on the use of water.

Total suspended solids (TSS): a measure of the nondissolved solid content in water, both natural and from waste disposal. Increased levels can affect drinking water appearance, clog irrigation equipment, or contribute to an increase in oxygen demand.

Time-weighted average: a measure computed by multiplying the number of days in a sampling period by the concentrations of individual constituents and dividing by the total number of days. The measure represents the expected composition of water for a site with equal quantities of water from the stream each day for the year.

Turbidity: an optical property of water defined as the reduction in the intensity of a light beam passed through water caused by suspended materials. Increased turbidity is undesirable in drinking water and can damage fish life.

Water quality limited: TDWR classification indicating a water quality problem in a segment. Segments receive this classification if (a) standards have been violated twice in a year, (b) receiving waters do not meet quality standards even when the federal effluent standards are met, or (c) advanced municipal waste treatment is required to protect water quality. Deviations from standards due to natural causes alone will not warrant this classification.

Water table: the upper surface of groundwater or the level below which the soil is saturated with water.

Water quality criteria: scientific data evaluated to derive recommendations for characteristics of water for specific uses.

Water quality monitoring: measuring of water quality parameters often required periodically by standards and generally done according to prescribed scientific procedures.

Water quality standards: definitions of acceptable quality generally derived from criteria.

Withdrawal: water taken from a surface or groundwater source for offstream use.

Bibliography

Alba, Francisco. "Mexico's Northern Border: A Framework of Reference." *Natural Resources Journal* 22 (October 1982): 749-763.

American Water Works Association, Incorporated. *Water Quality and Treatment: A Handbook of Public Water Supplies.* 3rd edition. New York: McGraw-Hill, 1971. 646 pp.

Armstrong, E. Neal. "Anticipating Transboundary Water Needs and Issues in the Mexico-United States Border Region in the Rio Grande Basin." *Natural Resources Journal* 22 (October 1982): 877-906.

Bailey, Larry. *Water Quality Segment Report for Segment No. 2311-- Pecos River.* Austin: Texas Water Quality Board, December 1974. 24 pp.

Baker, Roger. *Ground Water Resources of the Lower Rio Grande Valley Area, Texas.* USGS Water-Supply Paper 1653. Washington, D.C.: U.S. Government Printing Office, 1964. 55 pp.

Bean, Frank D.; Allan G. King, and Jeffrey S. Passel. *The Number of Illegal Migrants of Mexican Origin in the United States: Sex Ratio-Based Estimates for 1980.* Paper No. 4.020. Austin: Texas Population Research Center, 1982. 25 pp.

Bernard Johnson, Incorporated. *Texas Department of Water Resources and West Texas Council of Governments: Water Quality Management Plan for the Rio Grande Basin.* Volume 2. Houston: Bernard Johnson, Inc, July 1978. 153 pp.

Black and Veatch, Consulting Engineers. *Nonpoint Source Pollutant Sampling Program.* Project No. 8708. Dallas: Lower Rio Grande Development Council, June 1981. 62 pp.

Black and Veatch, Consulting Engineers. *Toxic Substance Pollutant Sampling Program.* Project No. 8708. Dallas: Lower Rio Grande Development Council, June 1981. 203 pp.

Blackwell, Charles. *Special Report on Oil Seepage into the Pecos River near Iraan, Texas.* Report No. SR-1. Austin: Texas Water Quality Board, June 1974. 8 pp.

Bluntzer, Robert L. *Selected Water Well and Ground Water Chemical Analysis Data, Ciudad Juarez, Chihuahua, Mexico.* Limited Distribution Report. Austin: Texas Water Development Board, 1975. 31 pp.

Bones, Jim, Jr., and John Graves. "Big River." *Texas Monthly* 10 (June 1982): 116-127, 212, 214.

Bower, Blair T. "Appraisal of Approaches Used to Forecast Water Demands." Unpublished. Arlington, Virginia, February 1980. 103 pp.

Bradley, David. "Health Problems on Water Management." *Journal of Tropical Medicine and Hygiene* 73 (June 1977): 286-293.

Bradshaw, Benjamin Spencer. "Potential Labor Force, Supply, Replacement, and Migration of Mexican-American and Other Males in the Texas-Mexico Border Region." *International Migration Review* 10 (Spring 1976): 29-45.

Casbeer, Thomas J., and Warren Z. Trock. *A Study of Institutional Factors Affecting Water Resource Development in the Lower Rio Grande Valley.* Report No. TR-21. College Station, Texas: Water Resources Institute, September 1969. 160 pp.

Carillo, E. P., and H. D. Buckner. *Index of Surface-Water Stations in Texas, January 1982.* Open-file Report 82-269. Austin: U.S. Geological Survey, 1982. 20 pp.

Coastal Bend Council of Governments. *Plan Summary Report for the Lower Nueces Basin (San Antonio-Nueces and Nueces-Rio Grande Coastal Basins) Water Quality Management.* Report No. LP-150. Austin: Texas Department of Water Resources, June 1978. 278 pp.

Day, John C. *Managing the Lower Rio Grande. An Experience in International River Development.* Chicago: University of Chicago, 1970. 274 pp.

Eibenschutz, Roberto. "La Planeacion del Desarrollo Urbano." *Natural Resources Journal* 22 (October 1982): 797-803.

El Paso, City of, Department of Planning, Research, and Development. *A Demographic Analysis of El Paso City and County.* El Paso: City of El Paso, December 1980. 57 pp.

Espey, Huston and Associates, Incorporated. "Testing of Sediments from the Inland Portion of the Brazos Island-Harbor Channel." Austin (No date). 50 pp.

Feachem, Richard. "Infectious Disease Related to Water Supply and Excreta Disposal Facilities." *AMBIO* 6 (October 1977): 59.

Flugrath, Marvin W., and Eleanor S. Chitwood. *Water-Quality Records for Selected Reservoirs in Texas, 1976-77 Water Years.* Report No. 271. Austin: Texas Department of Water Resources, September 1982. 189 pp.

Friebele, Charlotte D. *Bibliography of United States Geological Survey Reports on the Geology and Water Resources of Texas, 1887-1974.* Report No. 20-75. Washington, D.C.: U.S. Geological Survey, October 1975. 174 pp.

Friebele, Charlotte D., and Herbert A. Wolff. *Annotated Bibliography of Texas Water Resource Reports of the Texas Water Development Board and the U.S. Geological Survey.* Report No. 199. Austin: Texas Water Development Board, February 1976. 153 pp.

Gordon, Peter, and Peter Theobald. "Migration and Spatial Development in the Republic of Mexico." *Journal of Developing Areas* 10 (January 1981): 239-250.

Granger, C.W.I. *Forecasting in Business and Economics.* New York: Academic Press, Inc., 1980. 226 pp.

Greenberg, Michael R.; Donald A. Kruekeberg, and Connie O. Michaelson. *Local Population and Employment Projection Techniques.* New Brunswick, New Jersey: Center for Urban Policy Research, 1978. 277 pp.

Greenwood, Michael J.; Jerry R. Ladman, and Barry S. Siegel. "Long-Term Trends in Migratory Behavior in a Developing Country: The Case of Mexico." *Demography* 18 (August 1981): 369-388.

Grozier, Richard; Harold W. Albert; James F. Blakey, and Charles H. Membree. *Water-Delivery and Low-Flow Studies. Pecos River, Texas. Quantity and Quality 1964 and 1965.* Report No. 22. Austin: Texas Water Development Board, May 1966. 21 pp.

Hansen, Niles. "Economic Growth Patterns in the Texas Borderlands." *Natural Resources Journal* 22 (October 1982): 805-821.

Hedderson, John. "The Population of Texas Counties Along the Mexico Border." *Natural Resources Journal* 22 (October 1982): 765-781.

Holtz, David, and Scott Sebastian, Holcomb Research Institute. *Municipal Water Systems: The Challenge for Urban Resource Management.* Bloomington, Indiana: Indiana University Press, 1978. 299 pp.

Howe, Charles; Clifford S. Russell; Robert A. Young, and William Vaughan. *Future Water Demands--The Impacts of Technological Change, Public Policies, and Changing Market Conditions on the Water Use Patterns of Selected Sectors of the United States Economy: 1970-1990.* Report No. NWC-EES-71-001. Arlington, Virginia: National Water Commission, March 1971. 114 pp.

Hammer, Thomas R. *Planning Methodologies for Analysis of Land Use/ Water Quality Relationships.* Contract No. 68-01-3551. Washington, D.C.: Environmental Protection Agency, October 1976. 241 pp.

International Boundary and Water Commission. *An Appraisal of Potential Rio Grande Channel Storage Dams in Hidalgo and Cameron counties, Texas for Water Conservation.* El Paso, June 1981. 240 pp.

International Boundary and Water Commission. *Flow of the Rio Grande and Related Data--from Elephant Butte Dam, New Mexico to the Gulf of Mexico.* Water Bulletin Nos. 1-50. El Paso, 1930-1980.

International Boundary and Water Commission. *Rio Grande Boundary Preservation, Environmental Statement.* Draft. El Paso, September 1978. 251 pp.

Jamail, Milton H., and Stephen P. Mumme. "The International Boundary and Water Commission as a Conflict Management Agency in the U.S.-Mexico Lands." *Social Science Journal* 19, no. 1 (January 1982): 45-62.

Jamail, Milton H., and Scott J. Ullery. *International Water Use Relations Along the Sonoran Desert Borderlands.* Arid Lands Resource Information Paper No. 14. Tucson, Arizona: University of Arizona Office of Arid Lands Studies, 1979. 139 pp.

James Veltman and Associates, Incorporated. *208 Socio-Economic Report.* Report No. 208. McAllen, Texas: Lower Rio Grande Development Council, January 1977. 150 pp.

Johnson, W. Corwin. "Texas Groundwater Law: A Survey and Some Proposals." *Natural Resources Journal* 22 (October 1982): 1017-1030.

Keely, Charles B. "Illegal Migration." *Scientific American* 246 (March 1982): 41-47.

Kelejian, Harry H., and Wallace E. Oates. *Introduction to Econometrics--Principles and Applications.* New York: Harper and Row, 1981. 347 pp.

Kindler, Janusz, and Blair T. Bower. "Modelling and Forecasting of Water Demands." Paper presented at the conference on "Application of Systems Analysis in Water Management," Budapest, Hungary, November 1978. 28 pp.

Kirkpatrick, J. S. *Intensive Surface Water Monitoring Survey for Segment 2305--Amistad Reservoir.* Report No. IMS-21. Austin: Texas Water Quality Board, 1975. 38 pp.

Kirkpatrick, J. S. *Intensive Surface Water Monitoring Survey for Segment 2312--Red Bluff Reservoir.* Report No. IMS-58. Austin: Texas Water Quality Board, July 1977. 33 pp.

Kneese, Allen V., and Blair T. Bower. *Managing Water Quality: Economics, Technology, Institutions.* Baltimore: Johns Hopkins Press, 1968.   328 pp.

Kunofsky, Judith. *Handbook on Population Projections--How They Are Made and How They Make Themselves Come True.* San Francisco: Sierra Club, January 1982. 153 pp.

Kunofsky, Judith, and Donald Forman. *A Progress Report: State Implementation of EPA Guidelines on the Use of Population Projections.* San Francisco: Sierra Club, September 1979. 150 pp.

Long, Larry H., U.S. Department of Commerce, Bureau of the Census. *Interregional Migration of the Poor: Some Recent Changes.* Series P-23, No. 73. Washington, D.C.: U.S. Government Printing Office, November 1978. 29 pp.

Lower Rio Grande Valley Development Council. *Comprehensive Plan for Area-wide Sewerage Treatment Facilities, Vol. 1. Existing Conditions, Sewerage Systems, Water Quality.* McAllen, Texas, 1971. 281 pp.

Lower Rio Grande Valley Development Council. *Comprehensive Plan for Area-wide Sewerage Treatment Facilities, Vol. 2. Population Projections, Industrial and Domestic Waste Loads.* McAllen, Texas, 1971. 234 pp.

Lower Rio Grande Valley Development Council. *Comprehensive Plan for Area-wide Sewerage Treatment Facilities, Vol. 3.   The Plan.* McAllen, Texas, 1971. 301 pp.

Lyndon B. Johnson School of Public Affairs. *Guide to Texas State Agencies.* Fifth Edition.   Austin: University of Texas at Austin, 1978. 317 pp.

Mascorro, Edmundo. "Características fel Desarrollo Económico de la Franja Fronteriza del Norte de México." *Natural Resources Journal* 22 (October 1982): 1119-1123.

McBride, Robert H., editor. *Mexico and the United States: Energy, Trade, Investment, Immigration, Tourism.* Englewood Cliffs, New Jersey: Prentice-Hall, Inc., 1981. 197 pp.

McDonald, Brian, and John Tysseling. "Water Availability in the New Mexico Upper Rio Grande Basin to the Year 2000." *Natural Resources Journal* 22 (October 1982): 855-876.

McNeely, R. N.; V. P. Neimanis, and L. Dwyer. *Water Quality Sourcebook. A Guide to Water Quality Parameters.* Canada: Minister of Supply and Services, 1979. 88 pp.

Mendieta, H. B. *Reconnaissance of the Chemical Quality of Surface Waters of the Rio Grande Basin, Texas.* Report No. 180. Austin: Texas Water Development Board, March 1974. 109 pp.

Muller, Daniel A., and Robert D. Price. *Ground-Water Availability in Texas. Estimates and Projections through 2030.* Report No. 238. Austin: Texas Department of Water Resources, September 1979. 77 pp.

Mumme, Stephen P. "The Politics of Water Apportionment and Pollution Problems in United States-Mexico Relations." Unpublished. Department of Political Science, University of Arizona (No Date). 29 pp.

Mumme, Stephen P. "Regional Power in National Diplomacy: The Case of the U.S. Section of the International Boundary and Water Commission." Unpublished. Department of Political Science, Univerisity of Arizona, 1982. 39 pp.

Nanda, Ved P. *Water Needs for the Future. Political, Economic, Legal, and Technological Issues in a National and International Framework.* Boulder, Colorado: Westview Press, 1977. 329 pp.

New Mexico-Texas Bureau of Reclamation, Rio Grande Project Staff. *Water Resources of El Paso County, Texas.* Austin: Texas Water Development Board, 1973. 97 pp.

Office of the Governor, State of Texas. *Texas Trends: Texas 2000 Project.* HUD Sec 701. Austin: State of Texas, August 1980. 247 pp.

Ortiz, Alfonso. "Comparative Study on Infant Mortality in the Texas-Mexican Border Area of Laredo-Nuevo Laredo." Professional Report. University of Texas at Austin, 1983. 60 pp.

Ottmers, Donald. *Intensive Surface Water Monitoring Survey for Segment 2308, Rio Grande (Riverside Diversion Dam to New Mexico State Line)*. Report No. IMS-82. Austin: Texas Department of Water Resources, 1979. 19 pp.

Oyarzabel, Francisco. "Comentarios a las e Instituciones que Reglamentan las Aguas Superficiales de Mexico." *Natural Resources Journal* 22 (October 1982): 999-1005.

Oyarzabel, Francisco. "La Calidad de los Aguas del Bajo Rio Bravo." *Natural Resources Journal* 22 (October 1982): 925-937.

Plane, David A. "Where Do New Texans Come From?" *Texas Business Review* 56 (November-December 1982): 291-295.

Plauche, Breck, and Wlodzimierz Wojcik. *A Ground Water Manual for Small Communities*. Report No. CRWR-183. Austin: Center for Research in Water Resources, University of Texas at Austin, 1981. 245 pp.

Pound, Charles E.; Ronald W. Crites, and Douglas A. Griffes. *Costs of Wastewater Treatment by Land Application*. EPA-430/9-75-003. Washington, D.C.: U.S. Environmental Protection Agency, June 1975. 156 pp.

Prehn, W. Lawrence; Robert A. Sigafoos, and Ralph E. Childers. *The Economics of a Regional Municipal Desalting System in the Lower Rio Grande Valley of Texas*. Houston: Southwest Research Institute-- Houston, June 1967. 72 pp.

Rio Grande Compact Commission. *Budget Estimates FY 1984-1985*. Prepared for the Governor's Budget and Planning Office and the Legislative Budget Office. El Paso: Rio Grande Compact Commission, July 1982. 100 pp.

Rowe, Patricia M. *Country Demographic Profiles. Mexico*. Report No. ISP-DP-14. Washington, D.C.: U.S. Bureau of the Census, September 1979. 34 pp.

Safe Drinking Water Committee Advisory Center on Toxicology. "Drinking Water and Health." Chapters 6-7. Draft. Grant No. 68-01-3169. Washington, D.C.: National Academy of Sciences, 1977. 40 pp.

Secretaria de Agricultura y Recursos Hidraulicos. *Esquema de Desarrollo Hidrauliço para la Cuenca del Rio Bravo.* Mexico City: Secretaria de Agricultura Recursos y Hidraulicos, December 1980. 271 pp.

Secretaria de Agricultura y Recursos Hidraulicos. *Estudio de la Calidad del Agua en la Cuenca del Río Bravo.* Mexico City: Secretaria de Agricultura y Recursos Hidráulicos, 1978. 124 pp.

Secretaria de Recursos Hidraulicos--Dirección General de Distritos de Riego, Dirección de Estadística y Estudio Económicos. *Characteristicas de los Distritos de Riego Sigunda Edición Actualizada. Tomo I, Zonas Pacífico Norte, Norte Centro, y Noreste.* Mexico City: Secretaria de Recursos Hidráulicos, 1969. 184 pp.

Senate Regional Councils on Water Resources, Staff of. *Texas Water Administration: A Summary of the Governmental Units in Texas with Authority Over Water.* Austin: State of Texas, May 1974. 87 pp.

Sepulveda, Cesar. "Los Recursos Hidráulicos en la Zonia Fronteriza México-Estados Unidos Perspectiva de la Problematica Hacia el Año 2000--Algunas Recomendaciones." *Natural Resources Journal* 22 (October 1982): 1081-1092.

Siegel, Jacob S.; Jeffrey S. Passel, and J. Gregory Robinson. *U.S. Immigration Policy and the National Interest.* Appendix E to the Staff Report of the Select Commission on Immigration and Refugee Policy. Washington, D.C.: U.S. Government Printing Office, March 1981. 734 pp.

Sigler, Weston, and Winston Greenwood. *208 Nonpoint Source.* Vol. 1, Part 4. McAllen, Texas: Lower Rio Grande Valley Development Council, October 1978. 80 pp.

Slogget, Gordon. *Prospects for Ground-Water Irrigation. Declining Levels and Rising Costs.* Report No. 478. Washington, D.C.: U.S. Department of Agriculture, December 1981. 44 pp.

Straam Engineers, Incorporated. *Water Data Base Report.* EPA #P-0069 41-01-2. Uvalde, Texas: Lower Rio Grande Development Council, 1970. 101 pp.

Swenson, M. A., and H. L. Baldwin. *A Primer on Water Quality.* Washington, D.C.: U.S. Environmental Protection Agency, 1965. 27 pp.

Teclaff, Ludwick A., and Albert E. Utton, editors. *International Groundwater Law.* New York: Oceana Publications, 1981. 490 pp.

TerEco Corporation. *Environmental Baseline Description.* Report No. 208. McAllen, Texas: Lower Rio Grande Development Council, May 1977. 150 pp.

Texas A & M University Research Center. *Proceedings. The Inter-American Conference on Salinity and Water Management Technology.* El Paso: Texas A & M University Research Center, 1979. 221 pp.

Texas Department of Water Resources. *Budget Estimates, Fiscal Years 1984 and 1985.* First Submission. Austin, July 1982. 278 pp.

Texas Department of Water Resources, Planning and Development Division, Economics Section. "Fresh Water Use in Texas: 1974 and 1977." Computer Printout. Austin, 1982.

Texas Department of Water Resources. *Inventories of Irrigation in Texas, 1958, 1964, 1969, 1974, and 1979.* Report No. 263. Austin, October 1981. 295 pp.

Texas Department of Water Resources. *Ground-Water Development in the El Paso Region, Texas, With Emphasis on the Lower El Paso Valley.* Report No. 246. Austin, June 1980. 346 pp.

Texas Department of Water Resources. "Manufacturing Water Demand: 1974 and 1980." Computer Printout. Austin, 1982.

Texas Department of Water Resources. "Methods for Projecting Population for Texas Counties, 1990, 2000, 2010, 2020, and 2038." Draft. Austin, 1982. 18 pp.

Texas Department of Water Resources. "Report of Findings: Public Input to Amend the Texas Water Plan." Draft. Prepared for the Governor's Water Task Force, and the Natural Resources Advisory Council. Austin, June 1982. 133 pp.

Texas Department of Water Resources. "Self-Reporting Raw Data Report, Numeric Listing for May 21, 1982." Microfiche. Austin, 1982.

Texas Department of Water Resources, Enforcement and Field Operations. "Self-Reporting System Wasteload Data Report." Computer Printout. Austin, 1982.

Texas Department of Water Resources. *The State Of Texas Water Quality Inventory*. Report No. LP-59. Austin: Texas Department of Water Resources, 1980. 540 pp.

Texas Department of Water Resources, Planning and Development Division. "Texas Industrial Water Use Long Term Projections." Draft Two. Austin (No date). 108 pp.

Texas Department of Water Resources. *Texas Surface Water Quality Standards*. LP-71. Austin: Texas Department of Water Resources, April 1981. 107 pp.

Texas Department of Water Resources. "Wastewater Facility Needs, Middle Rio Grande Basin." Draft. Austin, April 1981. 8 pp.

Texas Department of Water Resources. "Water Planning Projections for Texas--1980-2030. Reported and Estimated Municipal, Manufacturing, Agricultural, Steam-Electric Power, and Mining Water Use in 1980; With Projections of Future Water Requirements at Different Growth Rates in 1970, 2000, 2010, 2020, and 2030." Review Draft. Austin, June 1982. 384 pp.

Texas Department of Water Resources, and West Texas Council of Governments. *Water Quality Management Plan for the Upper Rio Grande Basin*. Vol. 1. Basic Data Report. El Paso: West Texas Council of Governments, November 1977. 220 pp.

Texas Water Development Board. *Annotated Bibliography of Texas Water Resources Reports of the TWDB and USGS through August 1974*. Report No. 199. Austin, February 1976. 156 pp.

Texas Water Development Board. "Continuing Water Resources Planning and Development for Texas - Phase I." Draft. Vol. 1. Austin, May 1977. 59 pp.

Texas Water Development Board. "Continuing Water Resources Planning and Development for Texas - Phase I." Draft. Vol. 2. Austin, May 1977. 800 pp.

Texas Water Development Board. *Irrigation Benefit and Impact Analysis of the Trans-Pecos Area.* Austin, September 1971, 105 pp.

Texas Water Development Board. *A Summary of the Preliminary Plan for Proposed Water Resources Development in the Rio Grande Basin.* Austin, August 1966. 84 pp.

Texas Water Development Board. *A Survey of the Subsurface Saline Water of Texas. Geologic Well Data--Gulf Coast.* Vol. 8. Report No. 157. Austin, June 1980. 334 pp.

Texas Water Quality Board. *Existing Land Use Maps.* Atlas. Austin: Texas Department of Water Resources, February 1977.

Texas Water Quality Board. *Projected Land Use Maps: Year 2000, Rio Grande Basin.* Atlas. Report No. LP-56. Austin: Texas Department of Water Resources, February 1977.

Texas Water Resources Institute. *Directory of Water Resource Agencies and Organizations in Texas.* College Station, Texas: Texas Water Resources Institute (No date). 49 pp.

United Nations, Department of Economic and Social Affairs. *The Demand for Water: Procedures and Methodologies for Projecting Water Demands in the Context of Regional and National Planning.* Project ST/ESA/38. New York: United Nations, 1976. 240 pp.

United Nations, Department of Economic and Social Affairs. *Ground-Water Storage and Artificial Recharge.* Natural Resources/ Water Series No. 2. Report No. E.74.II.11. New York: United Nations, 1975. 270 pp.

U.S. Army Corps of Engineers. "Lower Rio Grande Basin, Texas. Flood Control and Major Drainage Project." Draft. Phase 1. Galveston, October 1980. 323 pp.

U.S. Bureau of Reclamation. *Texas.* Report No. 778-414. Washington, D.C., 1981. 25 pp.

U.S. Bureau of the Census. "Census of Population and Housing, 1980." Summary Tape File 3A, Texas. Computer Printout. Washington, D.C., 1980. 12 pp.

U.S. Bureau of the Census. *U.S.A. Statistics in Brief 1981.* Washington, D.C., 1982. 9 pp.

U.S. Comptroller General. *Better Data Collection and Planning is Needed to Justify Advanced Waste Water Treatment Construction.* Washington, D.C.: General Accounting Office, December 1976. 70 pp.

U.S. Comptroller General. *Report to the Congress, An Executive Summary: 16 Air and Water Pollution Issues Facing the Nation.* CED-78-148A. Washington, D.C.: General Accounting Office, October 1978. 43 pp.

U.S. Comptroller General. *16 Air and Water Pollution Issues Facing the Nation.* Appendix to the Report to the Congress. CED-78-148C. Washington, D.C.: General Accounting Office, October 1978. 107 pp.

U.S. Department of Agriculture, Texas Department of Agriculture. *1978 Texas County Statistics.* Austin: Texas Crop and Livestock Reporting Service, 1979. 507 pp.

U.S. Department of Agriculture. *Erosion and Sediment Control Guidelines for Developing Areas in Texas.* Temple, Texas: U.S. Department of Agriculture, 1976. 352 pp.

U.S. Department of Housing and Urban Development. "Areawide Environmental Impact Statement. Northwest Growth Area. Brownsville, Texas." Draft. San Antonio, September 1982. 411 pp.

U.S. Department of the Interior and U.S. Department of Agriculture. *Irrigation Water Use and Management.* Washington, D.C.: U.S. Government Printing Office, June 1979. 133 pp.

U.S. Department of the Interior. *Index of Surface Water Stations in Texas Operated by the Geological Survey in Cooperation with State and Federal Agencies.* Austin: U.S. Geological Survey, January 1981. 17 pp.

U.S. Department of the Interior. *Water Resources Data for Texas-Colorado River Basin, Lavaca River Basin, Guadalupe River Basin, Nueces River Basin, Rio Grande Basin.* Report No. TX-79-3. Austin: U.S. Geological Survey, 1979. 619 pp.

U.S. Department of the Interior. *Water Resources Data for Texas-Colorado River Basin, Lavaca River Basin, Guadalupe River Basin, Rio Grande Basin, and Intervening Coastal Basins.* Vol. 3. Report No. TX-80-3. Austin: U.S. Geological Survey, 1980. 583 pp.

U.S. Department of Justice, Immigration and Naturalization Service. *1978 Annual Report of the Immigration and Naturalization Service.* Washington, D.C.: U.S. Government Printing Office. 1979. 54 pp.

U.S. Environmental Protection Agency. *Annotated Bibliography for Water Quality Management.* 4th Edition. Washington, D.C., May 1977. 59 pp.

U.S. Environmental Protection Agency. *Clean Water and the Land: Local Government's Role.* Washington, D.C., January 1977. 23 pp.

U.S. Environmental Protection Agency. *Guidelines for State and Area-wide Water Quality Management Program Development.* Chapter 5. Washington, D.C., November 1976. 30 pp.

U.S. Environmental Protection Agency. *A Guide to the Clean Water Act Amendments.* OPA 1 29/8. Washington, D.C., November 1978. 19 pp.

U.S. Environmental Protection Agency. *Chemical Reporting and Record-Keeping Authorities Under 15 Environmental and Consumer Acts.* EPA-560/3-78-001. Washington, D.C., December 15, 1978. 452 pp.

U.S. Environmental Protection Agency. *Management Agencies Handbook for Section 208 Areawide Waste Treatment Management.* Washington, D.C., September 1975. 37 pp.

U.S. Environmental Protection Agency. *Proceedings: National Conference on Water Conservation and Municipal Wastewater Flow Reduction.* Contract No. 68-03-2674. Rockville, Maryland: Enviro Control, Inc., November 1978. 288 pp.

U.S. Environmental Protection Agency, Office of Water Planning and Standards. *Quality Criteria for Water.* Washington, D.C., July 1976. 256 pp.

U.S. Environmental Protection Agency. *Report on State Sediment Control Institutes Program.* Report No. EPA-440/9-75-001. Washington, D.C., April 1975. 27 pp.

U.S. Environmental Protection Agency, Water Planning Division. "Alternative Growth Management Techniques." Draft. Washington, D.C., 1976. 103 pp.

U.S. Environmental Protection Agency. *Water Quality Criteria 1972.* Report No. EPA R3-73-033. Washington, D.C.: National Academy of Sciences, March 1973. 223 pp.

U.S. Geological Survey. *Index of Surface Water Stations in Texas Operated by the Geological Survey in Cooperation with State and Federal Agencies.* Austin, January 1981. 53 pp.

U.S. Geological Survey. *Reconnaissance Investigations of the Ground-Water Resources of the Rio Grande Basin, Texas.* Bulletin No. 6502. Second Printing. Austin: Texas Water Commission, August 1973. 213 pp.

U.S. Geological Survey. *Water Resources Data for Texas, Water Year 1978, Volume 3--Colorado River, Lavaca River, Guadalupe River, Nueces River, Rio Grande Basins and Intervening Coastal Basins.* Report No. USGS/WRD/HD-80/002. Austin, October 1979. 607 pp.

U.S. Geological Survey. *Water Resources Data for Texas, Water Year 1979, Volume 3--Colorado River, Lavaca River, Guadalupe River, Nueces River, Rio Grande Basins and Intervening Coastal Basins.* Report No. USGS/WRD/HE-80/005. Austin, September 1980. 619 pp.

U.S. Water Resources Council. "The Nation's Water Resources, The Second National Water Assessment by the U.S. Water Resources Council, Lower Colorado Region (15)." Draft. Washington, D.C., April 1978. 89 pp.

U.S. Water Resources Council. *The Nation's Water Resources 1975-2000. Vol. 4: Rio Grande Basin.* Washington, D.C.: U.S. Government Printing Office, December 1978. 61 pp.

U.S. Water Resources Council. *The Nation's Water Resources 1975-2000. The Second National Water Assessment by the U.S. Water Resources Council, Summary.* Washington, D.C.: U.S. Government Printing Office, January 1979. 86 pp.

U.S. Water Resources Council. "The Nation's Water Resources. The Second National Water Assessment by the U.S. Water Resources Council, Arkansas-White-Red Region (11)." Draft. Washington, D.C., April 1978. 140 pp.

U.S. Water Resources Council. *The Nation's Water Resources 1975-2000. Vol. 2: Water Quantity, Quality and Related Land Considerations.* Washington, D.C.: U.S. Government Printing Office, December 1978. 592 pp.

U.S. Water Resources Council. *Standards for Planning Water and Land Resources.* US-WRC-0092. Washington, D.C.: National Information Service, July 1970. 296 pp.

Urban Engineering. *Comprehensive Plan for Municipal, Water, Sanitary Sewerage and Storm Drainage Facilities.* Middle Rio Grande Development Council, July 1978. 100 pp.

Utton, Albert E. *Selected Resource Issues in the Border Region: Anticipating Transboundary Resource Needs and Issues in the U.S. Mexico Border Region.* Albuquerque, New Mexico: Natural Resources Center, University of New Mexico (No date). 22 pp.

Valdes, Carlos Rincon. "De Aguas Subterraneas en la Región de Juarez-El Paso." *Natural Resources Journal* 22 (October 1982): 847-853.

Warner, Dennis, and Jarir S. Dajani. *Water and Sewer Development in Rural America: A Study of Community Impacts.* Contract No. 30006. Lexington, Massachusetts: Lexington Books, 1975. 128 pp.

Warshaw, Steve. *Water Quality Segment Report for Segment No. 2302--Rio Grande.* WQS-15. Austin: Texas Water Quality Board, June 1975. 23 pp.

Warshaw, Steve. *Water Quality Segment Report for Segment No. 2306--Rio Grande.* WQS-9. Austin: Texas Water Quality Board, February 1975. 25 pp.

West Texas Council of Governments. "Interim Report: Wastewater Facility Needs; Upper Rio Grande Basin." Draft. El Paso, January 1981. 49 pp.

West Texas Council of Governments. *Plan Summary Report for the Rio Grande Basin Water Quality Management Plan.* Report No. LP-166. Austin: Texas Department of Water Resources, June 1981. 175 pp.

West Texas Council of Governments. *Water Quality Management Plan for the Upper Rio Grande Basin.* Basic Data Report. El Paso: West Texas Council of Governments, November 1977. 233 pp.

Whetstone, George A. *Re-Use of Effluent in the Future, with Annotated Bibliography.* Report No. 8. Austin: Texas Water Development Board, December 1965. 187 pp.

White, Donald; Joseph Gates, James Smith, and Bennie Fry. *Ground-Water Data for the Salt Basin, Eagle Flat, Red Light Draw, Green River Valley, and Presidio Bolson in Westernmost Texas.* Report No. 259. Austin: Texas Department of Water Resources, October 1980. 97 pp.

Wilson, David L., and Harry W. Ayer. *The Cost of Water in Western Agriculture.* Report No. AGES830706. Washington, D.C.: U.S. Department of Agriculture, August 1982. 16 pp.

Wilson, Leonard U. *State Water Policy Issues.* RM-652. Lexington, Kentucky: Council of State Governments, November 1978. 63 pp.

Whittington, Dale. *Forecasting Industrial Water Use.* RM-78-71. Laxenburg, Austria: International Institute for Applied Systems Analysis, December 1978. 72 pp.

World Bank. *1978 World Bank Atlas: Population Per Capita Product and Growth Rates.* Washington, D.C.: World Bank, 1978. 32 pp.

Index

325

71227